Raising Healthy Poultry

by

Dr. W. Malcolm Reid, Dr. Gene M. Pesti,
Dr. Billy Hargis, Dr. Randle Moore,
Dr. Pran Vohra, Dr. William F. Dean,
Dr. M. A. Hammarlund

Christian Veterinary Mission

Christian Veterinary Mission
19303 Fremont Avenue North
Seattle, WA 98133 USA
Current Book Information at:
www.cvmusa.org
PHONE: (206) 546-7569 FAX: (206) 546-7458

Permission requests should be addressed to: vetbooks@cvmusa.org

OR

Christian Veterinary Mission
19303 Fremont Ave N
Seattle, WA 98133 U.S.A

THIS IS THE FOURTH EDITION

It can be improved with your help. If you are a veterinarian, veterinary technician, poultry producer, animal husbandry worker, missionary, development worker or anyone with ideas or suggestions for ways this booklet could be changed to better meet the needs of your people, please write to the authors at:

Christian Veterinary Mission
19303 Freemont Avenue North
Seattle, WA 98133 USA
FAX 206/546-7458
E-mail: vetbooks@cvmusa.org
Web site: www.cvmusa.org

Thank you for your help.

Library of Congress Cataloging-in-Publication Data

Raising healthy poultry / by W. Malcolm Reid ... [et al.].
p. cm.
Includes bibliographical references.
ISBN 1-886532-02-8
1. Poultry. 2. Poultry—Nutrition. 3. Poultry—Diseases. 4. Poultry—Health. I. Reid, W. Malcolm.
SF487.R25 2006
36.5'0852—dc22

2005037830

Since the original poultry book was published in February 1990, the author, W. Malcolm Reid, died in November 1990. This book has been so well received that we are printing the third edition and second revision of the book. Malcolm's original work continues to be an effective and practical book. We offer a prayer in his memory.

Prayer

Our heavenly Father, we thank you for the talents of Malcolm Reid and for his willingness to share those talents with your children in need around the world. We thank you for the help that this book has brought to so many. We pray your continued blessings on this material, that it will serve in a very practical way to help those who have an interest and need to be involved in poultry production. We pray that it will also point to the hope we have in your Son, Jesus Christ and in whom Malcolm had his faith. Amen.

Raising Healthy Animals Series

Every year, thousands of people around the world struggle to survive because they don't have the right knowledge, skills and resources to care for their animals. Christian Veterinary Mission (CVM) sends veterinary professionals to live and work alongside many of these people to encourage them and provide them with not only much needed veterinary expertise, but also the hope that is only found in Christ. CVM veterinarians build lasting relationships with individuals and communities, helping them be transformed through Christ's love.

CVM, in its effort to be meaningfully involved in work in the developing world, quickly found there was little appropriate educational material available. CVM set about developing basic resource materials in animal husbandry for farmers and agricultural workers. Apparently, they met a real need, as these books have been accepted throughout the developing nations of the world.

The series of books published by Christian Veterinary Mission includes the following in order of publication:

Raising Healthy Pigs *
Raising Healthy Rabbits *
Raising Healthy Fish
Raising Healthy Cattle
Raising Healthy Poultry *+
Raising Healthy Goats *
Raising Healthy Sheep
Drugs and Their Usage
Where There Is No Animal Doctor
Raising Healthy Horses
Zoonoses: Animal Diseases That Affect Humans
Raising Healthy Honey Bees
Slaughter and Preservation of Meat
Disease and Parasite Prevention in Farm Animals

[Also available in: * Spanish + French].

CVM fieldworkers have also developed specific training materials for the countries in which they work.

All of these books have been put together by Christian men and women; in a labor of love and service, for people in need throughout the world. It demonstrates dedication to their profession, service to humanity and a witness to their faith. We hope that they are a help to you in developing an appropriate livestock program to meet your needs. We pray God's blessing on their use.

Leroy Dorminy
CVM Founder

TABLE OF CONTENTS

SECTION 7—Poultry Health Programs

SECTION 8—Marketing

SECTION 9—Ducks

SECTION 10—Other Kinds of Poultry

SECTION 11—Other Information

Section 1

Introduction

More people raise poultry than any other food-producing animal. They may begin as a hobby for pleasure, to produce their own food, or for profit. This booklet may be regarded as a primer on essential methods of keeping poultry healthy. Suggestions are made on how to expand into a small-scale poultry farm under primitive conditions.

Since chickens are the most common domesticated bird raised on farms, villages and even in towns or cities, they will be used to illustrate poultry management which may also apply with some modifications for ducks, geese, turkeys, guinea fowl, quail, and pigeons. Some contrasts between chickens and other species are made in Section 10.

Many poultry producers have started by rearing a few birds as pets. Associated income, which was at first regarded as "pin-money", may grow into plans for a larger commercial enterprise. Optimistic hopes for financial returns may mislead the producer into too rapid expansion if careful planning has not taken place.

The best advice: "Start small the first year and let financial success govern your rate of expansion. Keep careful records!" Many financial failures have come from attempts to start with large flocks without knowing the best sources and costs of high-producing chicks, feed, methods of disease prevention, or marketing. This booklet may assist small-scale poultry operators as they expand into a successful commercial enterprise. Large scale poultry production is a specialized agricultural industry which will require further study of more detailed textbooks (See Section 11).

The potential of eggs and especially egg protein to help solve malnutrition problems common to children in the underdeveloped areas of the world is well known. Children between one and three years of age are most vulnerable to low protein diets. Protein deficient children may have irreversible physical and mental retardation, and become weakened and are highly susceptible to infectious diseases. Eggs can markedly improve the quality of their lives and their children's lives.

Section 2

Getting Started—Securing Stock

Which one shall I order?

Chickens are conveniently purchased and moved to new locations at four different stages in their lives:

1) As day-old baby chicks. This is the most common method of starting or restocking a poultry operation. Advantages: Since day-old chicks are nourished by stored yolk for 2 or 3 days, they do not require food or water during shipment to new quarters. Disadvantages: during shipment there is danger of smothering if overheated or getting chilled in freezing temperature. Although newly hatched baby chicks require more heat than humans, groups of 25 generate sufficient heat to be comfortable in a cardboard shipping box if some ventilation holes are provided and moderate room temperatures can be maintained. Watering and feeding will be necessary if shipments are delayed more than 3 days.

2) Purchase of hatching eggs secured from good breeding stock makes another convenient method of getting started. Although eggs may be easily broken if insecurely packed, temperature changes are less critical than with baby chicks. The producer must be prepared to incubate the eggs as soon as they are received. Hatching eggs cannot be stored for much longer than a week even if maintained at the optimum temperature of 10°C (50°F).

Natural incubation requires availability of a "broody" hen, turkey, duck or other bird (see the section on breeding practices). Although numbers of eggs set under a hen may be limited to 12-14, other birds sometimes incubate larger numbers of eggs. Hatching still larger numbers of chicks requires use of a mechanical incubator for *artificial incubation.* Success with this method is dependent upon skill and care of a the hatchery operator. If eggs from disease-free stock are obtained, care must be taken to prevent contamination by eggs or chicks from other sources. Percentage of hatch depends upon egg fertility, nutrition of the hen, condition of the egg shell, genetic constitution of the parents, egg storage time and temperature before incubation.

3) Purchase of started chicks three weeks or more of age saves the beginner from supplying equipment and the time-consuming tasks associated with *brooding.* These include use of artificial heat, initiation of chicks into feeding and drinking and sometimes early vaccinations.

4) Purchase of started pullets ready to lay at about 20 weeks of age relieves the customer of numerous rearing tasks. Advantages: males and inferior females are eliminated before purchase. Males are often destroyed as day-old chicks in the hatchery after *sexers* have examined the vent under strong light or by noting genetic feathering differences inherited from selected parent lines. Vaccination and other disease prevention procedures may have been accomplished. Disadvantage: added costs are usually charged an a per bird basis for these services.

Section 3

Types of Chickens and Breeding Practices

Poultry producers generally make a choice in selecting their stock from three types of birds:

Dual-purpose chickens are reared to produce both eggs and meat. Most young males are sacrificed after 2 or 3 months for meat while females are maintained for a year or more to produce eggs. Females are also salvaged for meat after egg production declines. Dual-purpose birds are generally preferred for backyard flocks, while larger commercial poultry producers generally specialize in either meat or eggs. A line of chickens developed especially for non-intensive small-scale poultry producers is the *"Triple Production Reds"* developed by Dr. John P. Bishop of 11806 State Route 347, Marysville, OH 43040 USA (telephone 937/348-2344; E-mail: *kbishop@urec.net*). This brown egg strain will go broody, incubate eggs and care for chicks.

Egg-layer strains have been selected for good egg production. Surplus males are often eliminated on hatch-day after sexing since they are usually genetically inferior for meat production. Spent females (at end of lay) may be salvaged for meat after their egg production drops below 50%. Their salvage value usually is minimal because of their small size.

Meat chickens, often known as *broilers* or *fryers,* are reared from strains selected for rapid growth. Both male and female lines are selected from parent stock which produce rapid meat production. Some of these hybrid strains produce a bird weighing 2 kilos (4.4 lbs.) in 7 weeks after being fed less than 4 kilos (8.8 lbs.) of a well-balanced diet.

Primary breeding companies with hybrids. Specialized breeding organizations have developed feed-efficient strains of these three types of chickens. They are produced by cross-breeding techniques and go by the general name of *"hybrids".* This cross breeding brings about gene combinations which excel in production of eggs or meat over that of either parent. Superior qualities produced by new gene combinations are attributed to *"hybrid vigor".* These hybrid strains are now so popular with modern poultry producers that they have abandoned use of such dual-purpose breeds as Rhode Island Reds, New Hampshire, Cornish, or Barred Plymouth Rocks. Improved hybrid strains of White Leghorns produce white eggs while brown or tinted egg strains usually have some White Leghorn ancestry.

These breeding companies require years of work to test dozens of strains and hundreds of progeny. Certain inbred or synthetic strains which combine well are then selected to become the great-grandparents or grandparents of chicks sold to poultry producers. Strain crosses known as the male and female lines are established and sold separately to hatchery organizations which rear them together as their "breeder flocks". All hatching eggs are derived from such breeders.

Poultry producers purchasing these hybrid chicks have at first been reluctant to pay for the necessary increased cost for these hybrid chicks. However, the cost of the hybrids is more than compensated for, if the purchase price of feed consumed is also calculated. Unfortunately this hybrid vigor is rapidly lost by interbreeding the hybrid birds. The producer needs to renew the male and female lines of breeder stock in each generation.

Names of many of these hybrid strains are now recognized by poultry producers on a world-wide basis. Broiler strains are often named for the breeder organizations such as Arbor Acres, Cobb, H & N, Hubbard, Indian River, Peterson, Pilch, Ross, Shaver, Vantress, and Vedette. Some of the egg-type strains have been produced by Babcock, HyLine, DeKalb, Hisex, and Tatum. Names of some dual-purpose strains would include: DeKalb-Warren 6-Sal-Link (USA), Hardy sex-linked (USA), Kabir (Israel), Label (France), Nera sex-linked (Japan), Parks Hybrids, and Stino's White Baladi (Egypt).

Dual-purpose breeds are still preferred in many parts of the world where brooding hens are used to hatch and start chicks. A running debate has continued in many countries on whether to select native breeds or imported hybridized stock. Advantages sometimes cited for native breeds include: disease resistance inherited from long natural exposure, and local customer preference for these more "tasty" and "chewy" birds. Differences in taste may be due to scavenged feedstuffs such as garlic, insects, herbs, and manure. Slow growth probably produces the more "chewy" meat. National pride has sometimes influenced local decisions. Imported disease-free hybrids may grow faster, have better feed conversion, and produce more eggs. Natural inbreeding of backyard flocks results in low productivity for both meat and eggs.

By introducing imported males to breed with native hens, some of the advantages of hybrid vigor can be inexpensively incorporated in village-wide programs where producers are still dependent upon natural hatching and brooding. A happy "marriage" of improved egg production, growth and disease resistance from the males and local characteristics is thereby achieved. The program calls for elimination of all native cocks and substituting the surplus males coming from imported lines which have been selected for high productivity. Better egg and sometimes better meat production has been reported from such programs in Egypt, Ghana and India. The imported male chicks are inexpensive as they are otherwise often discarded in the hatchery.

Section 4

Housing and Equipment

A. Housing is considered essential to protect chickens from predators, rain, wind, and temperature extremes. Although the ancestral jungle fowls roost in trees at night, mortality losses are often high. Nocturnal (night time) attacks from dogs, cats, rats, owls, hawks and other predators may destroy even adults chickens. Protective housing is very desirable (Fig. 1-4). If properly constructed and managed the house may also protect against the spread of diseases and parasites. Satisfactory houses have been built from wood, stone, concrete, adobe, brick, bamboo, and sheet metals. Materials available and designs used vary so greatly from different areas that detailed plans may best be secured from local authorities. Several essentials need to be considered before building a house.

Fig. 1. Free range management.

Fig. 2. Night shelter for partially controlled range.

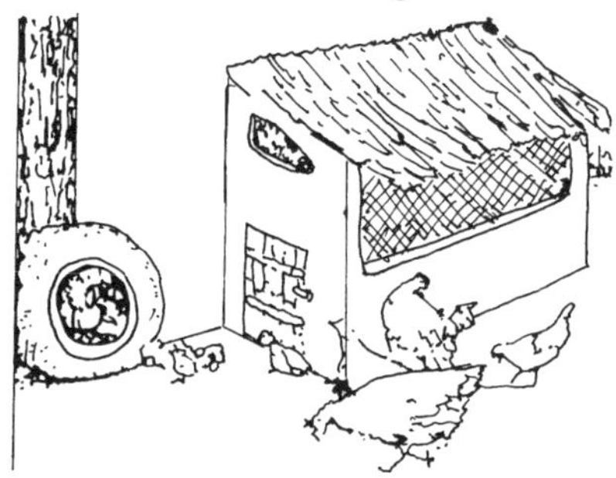

Fig. 3. Gable chicken house.

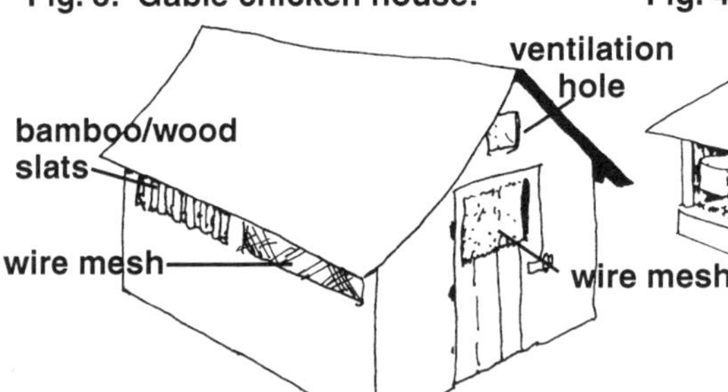

Fig. 4. Deep litter house for confinement rearing.

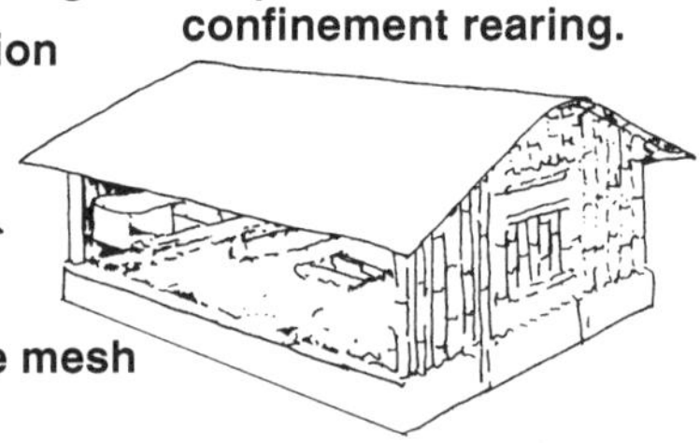

Location. The house needs to be properly oriented to 1) protect against wind but permit some air movement, 2) prevent too much heat from the sun, and 3) to insure sufficient drainage to prevent wet floors. In hot climates available shade and roof overhang may need consideration. In the northern hemisphere windows are often oriented to leave an open southern exposure. If built in deep valleys there may be insufficient air circulation. Objectionable odors or houseflies breeding in poultry houses have forced some poultry units to move from developing suburbs.

Temperature Control. Temperature extremes in hot and cold weather create stressful conditions in poultry houses. Over heating of drinking water may be dangerous in hot climates. Frozen waterers create special problems in cold weather. Specialized construction plans adapted to local conditions may be available from local experts.

Ventilation. Air movement within the house is essential to prevent smothering (Fig. 3). Chickens need more fresh air per unit of body weight than any other class of livestock. Wide open sides or windows may be desired in hot climates. Air movement should not be blocked by bushes or other buildings. Ventilators and/or fans may be essential in large commercial houses.

Insulation. Health of poultry is often promoted by insulating roofs and sidewalls against both cold and hot weather. Advice on materials and engineering recommendations should be sought locally before building large houses.

Moisture Control. Large quantities of water are consumed by chickens. Layers may require more than a liter per day each in hot climates. In humid climates high humidity may result in wet droppings which cause disease problems. Adequate drainage in the house floor may assist in moisture control. To prevent moisture problems some producers have selected management programs using slatted or wire floors.

Space. Crowded chickens are unhappy and unproductive. Minimum requirements for birds housed in a moderate climate with plenty of ventilation are as follows: 15 birds per square meter up to 6 weeks of age, 5 birds per square meter up to 16 weeks, and 3 to 4 birds per square meter for older birds. Higher temperatures, excess humidity, or wet litter increase required floor space requirements. In cages, 10 to 20 layers may be placed per square meter of floor space. Larger white and brown egg layers require proportionately greater floor space, as calculated by weight.

Framing. Availability of local materials and construction methods will determine the type of framing. Internal pillars may be required to support the roof against heavy wind or snow. Special strength is required over doors, windows and if tile roofing is used.

Roof. The roof should be entirely rainproof. Thatch, sheet metal, tile, shingles or wood covered by roll roofing materials may be used. An overhang of up to one meter on the south side (northern hemisphere) may provide shade in summer but permit sun to enter the house during winter. The amount of overhang to give this advantage varies with the latitude.

Floor. Although concrete floors with embedded wire mesh to keep rats out are desirable, more houses have dirt or heavy clay floors because of expense. Adequate drainage should be planned. Types of floors vary with different management systems which include deep litter, and suspended floors made of slats or wire. Convenient clean-out systems for manure need to be considered.

Animal and Bird Protection. Nocturnal (night time) protection against dogs, cats, rats, weasels, owls and day-time protection against wild animals, hawks and eagles should be planned. Wire mesh is often used. Although foraging in pasture (free range) may be satisfactory during the day, greater protection is usually required at night.

B. Waterers. *Chickens of any age should never be left without clean, cool water!* Chickens need water every 15 to 20 minutes. Making certain that water supplies are adequate is one of the most important jobs of the caretaker. Waterers adapted for various ages of chickens are illustrated (Fig. 5). The simplest waterer is a tin can inverted into a soup or pie plate (Fig. 6). Punch a hole about 1.5 cm (0.5 in.) from the open end of the tin can. Fill the can with water and cover it with the plate. With one hand on the plate and one on the tin can, quickly invert both. The position of the punched hole and the vacuum in the tin can will regulate the

Waterers

Fig. 5. Chick waterer.

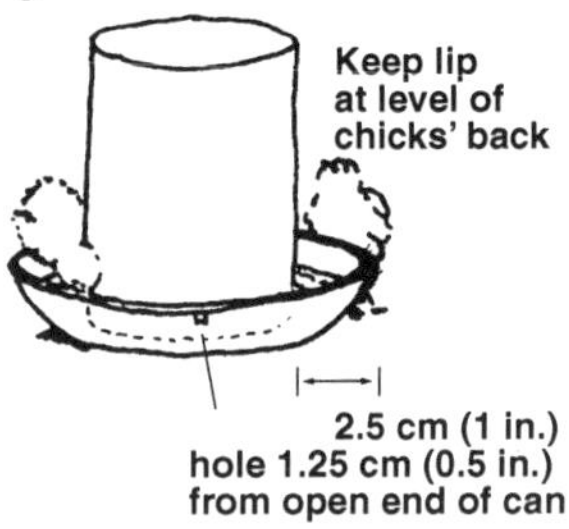

Fig. 6. Parts of waterer.

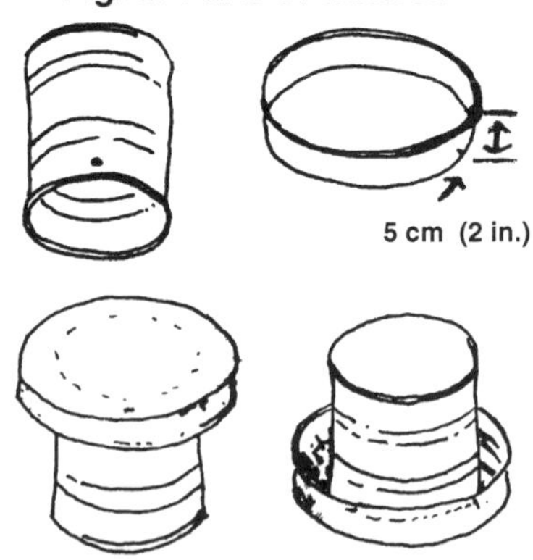

Fig. 7. Clay waterer.

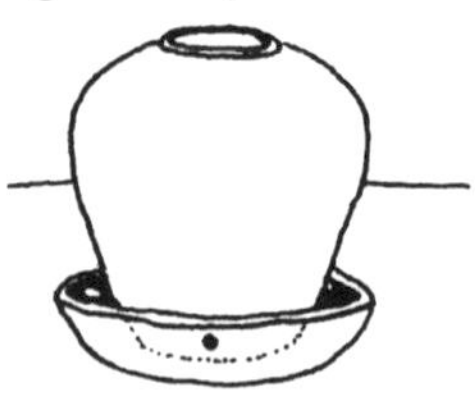

Fig. 8. Bottle waterer.

Fig. 9. Gourd waterer.

Fig. 10. Slat holder.

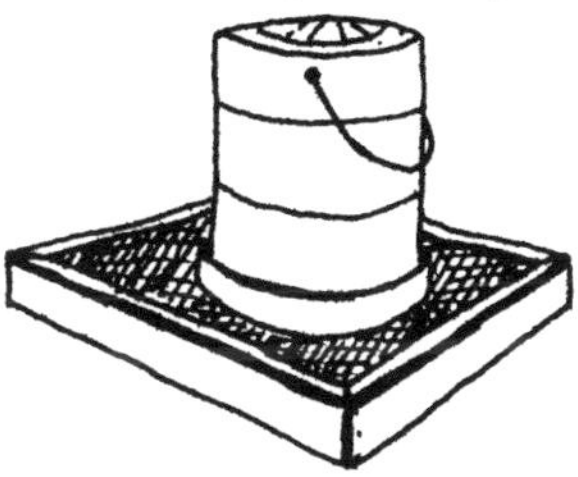

Fig. 11. Waterer placed to keep litter dry.

Fig. 12. Automatic waterer.

water level in the plate. Smaller cans are suitable for chicks but larger ones must be substituted as growing birds require more water.

Other waterers can be fashioned using a clay pot and shallow plate with a hole or groove in the lip (Fig. 7). A removable bottle strapped to the wall or a stand can be adjusted at a convenient height above a dish (Fig. 8). A gourd submerged in the ground for stability has holes of convenient size cut in the sides (Fig. 9).

Various automatic devices have been invented to preclude refilling waterers so frequently (Fig. 12). Illustrated is one built with a clean 50 gallon oil drum connected to a galvanized sheet metal trough 10 cm (4 in.) high. Flow of inlet water into the drinker is adjusted by a faucet and an overflow pipe leading outside the house. The top of the waterer needs to be protected by a spinner (a bar that rotates so chickens cannot roost on it). All waterers need frequent checking to see that water is constantly available but that no overflow goes into the litter. To keep the litter dry, each waterer is often placed on a flat board between slats (Fig. 10) or suspended on a wire-floored platform 9 cm (3 1/2 in.) above the litter (Fig. 11).

Slime molds begin to grow in waterers if they are not washed frequently. This should not be allowed to happen. A long handled brush is often used to clean out debris. Waterers need periodic cleaning with a brush. Some authorities suggest that this be done daily.

C. Feeders. Providing a continuous supply of feed for chickens is one of the major tasks of the poultry producer. Elongated trough-type feeders with overhanging lips on both sides may be constructed of wood, metal or bamboo (Figs. 13-20). Sufficient linear feeder space along the trough should be provided for all birds to feed at once. Otherwise timid birds may starve to death while others resort to cannibalism. See Fig 12A (taken from Pingel (1981) Kleintiere richtig füttern). To avoid feed wastage, feeder height should be continuously adjusted to be even with the height of the chicks' shoulders. The chicks or hens should have to exert the smallest amount of energy in reaching up and into the trough.

Feeders

Fig. 13. Chick feeder.

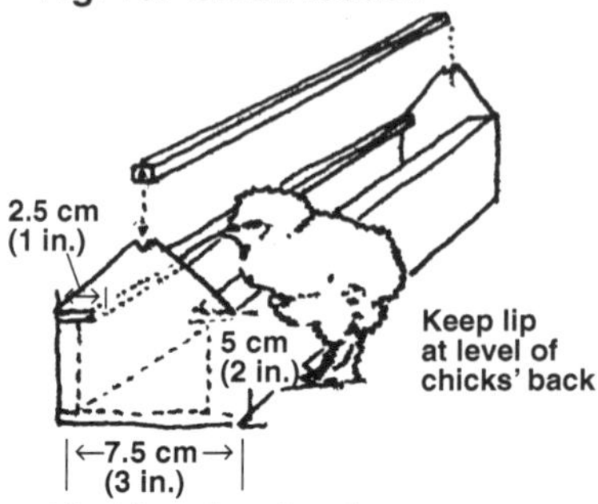

Fig. 14. Older chick feeder.

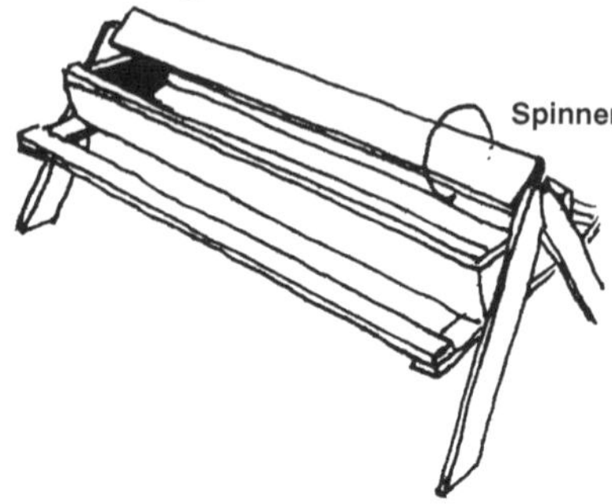

Fig. 15. Hanging feeder.

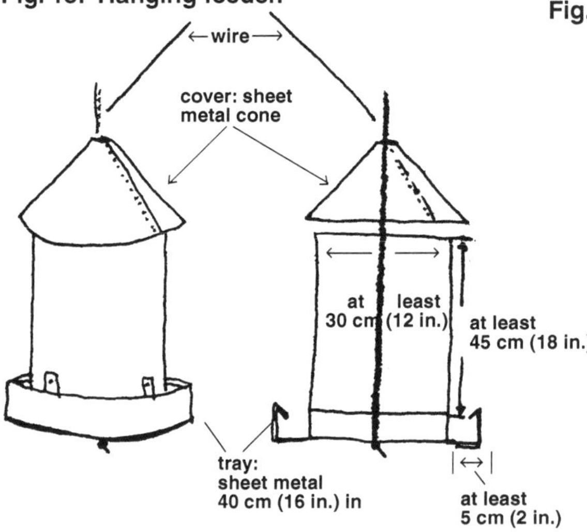

Fig. 16, 17, 18. Lip overhang.

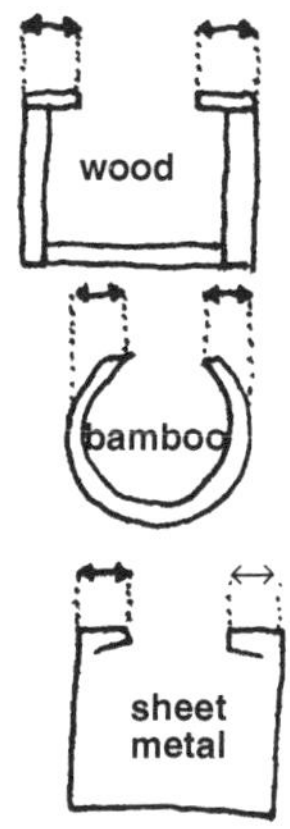

Fig. 19. Hanging feeder—bamboo.

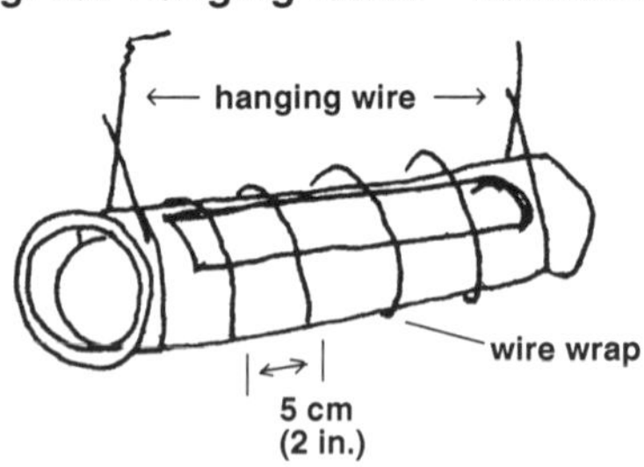

Fig. 20. Stationary feeder—bamboo.

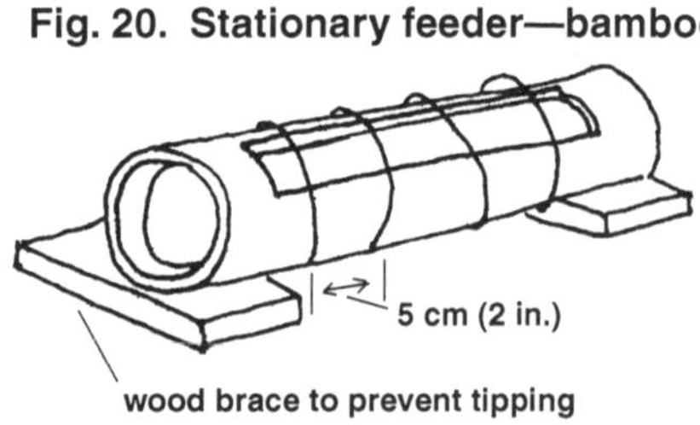

Diagram by Erik

D. Cages. Craftsmen can fabricate cages if supplied with welded wire mesh with wire 2 mm (1/8 in.) in diameter. For the frame and dividers wire size is spaced 2.5 x 5 cm (1 x 2 in.). A piece cut 3.66 m x 1.5 m (12 x 5 ft.) provides the form for 3 colony cages or 16 double cages (Figs. 21 and 22). Four dividers plus end pieces cut 40 x 45 cm (16 x 18 in.) will make three colony cages or 17 for 16 double cages spaced 22.8 cm (9 in.) apart. A flexible metal wire is used to bind the dividers to the frame. The front is made from wire mesh with larger openings (5 x 10 cm, 2 x 4 in.). Doors may be cut in as indicated in Fig. 23. Floors should slant downward 7.6 cm (3 in.) back to front so eggs will roll through a 5 cm (2 in.) gap in the front and rest upon the upturned floor for easy collection (Fig. 26). Bamboo or slats may be substituted for wire floors (Figs. 24, 25). Feeders and waterers may be made of sheet metal or bamboo and hung tightly to the front of the cage. Spacing is important to prevent spillage of water or feed. Double cages accommodate 2 layers each while colony cages permit 6 if they have had their beak trimmed.

A system of double hanging is illustrated (Fig. 27). A correct slant of 7.6 cm (3 in.) to the floor is required to permit eggs to roll out without damage (Fig. 26).

The "chicken tiller" in Figure 27B is an adaptation of the age-old practice of moving animals from one area to another to consume fresh forage and deliver nutrients to the soil. An added benefit of using chickens is that they will scratch the soil with their feet, and so incorporate, or till nutrients into the soil. The chickens in the enclosure are given feed and water for a few weeks before the "chicken tiller" is moved, and soil enrichment in a new location is begun.

The tilled and fertilized soil is then an ideal place to plant vegetables. A careful balance must be achieved between the number and size of the chickens, the size of the enclosure and the length of time the "chicken tiller" is kept in any one location. Although plants need some nitrogen to grow, too much nitrogen inhibits the germination of most seeds and otherwise stunts plant growth. Too little fertilizer will probably prove much better than too much, since more manure can be added as the plants grow. The "chicken tiller" in the drawing has been successfully used by the Heifer Project International, P. O. Box 808, Little Rock, AR 72203 (telephone 1-800-422-0474) (drawing by Cindy King, Heifer Project International). This practical application of using a few chickens to enhance soil fertility is most helpful. Confine a few chickens—provide food and water—after 6 or 8 weeks remove "Tiller" and plant vegetables. Repeat cycle.

Feeders

Fig. 21. Feeder and waterer location.

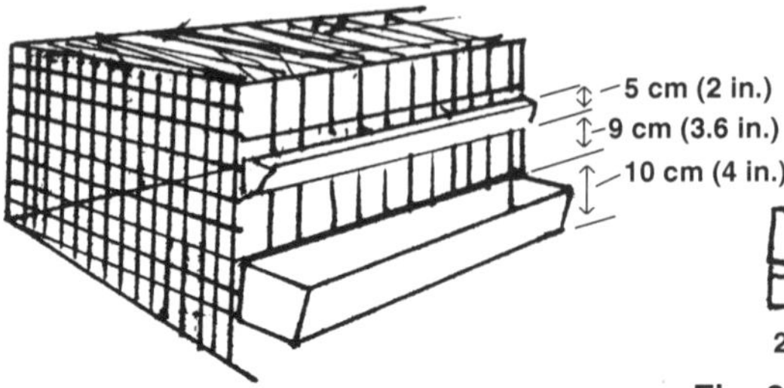

Fig. 24. Floors may be slatted.

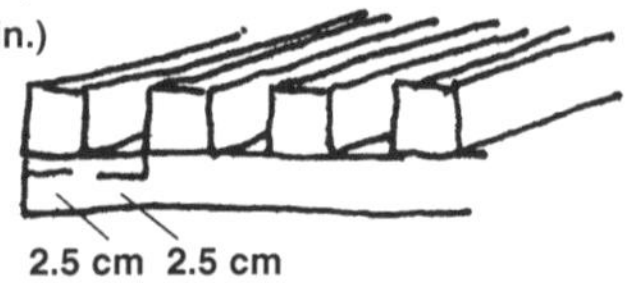

Fig. 25. Floors may be bamboo.

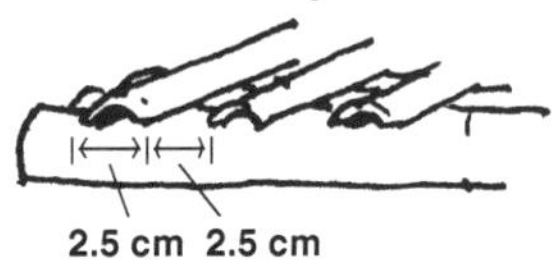

Fig. 22. Form for cage from wire mesh.

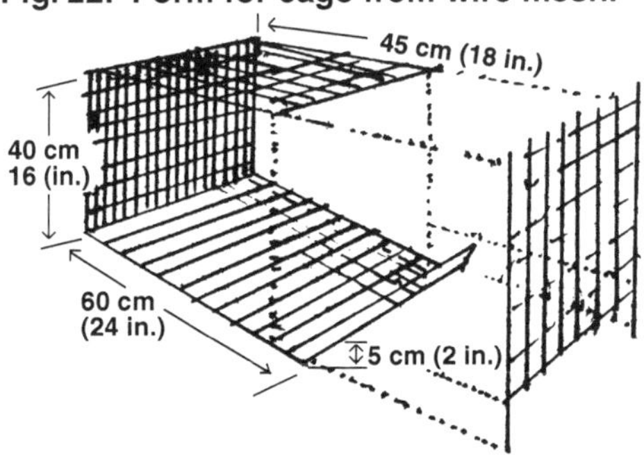

Fig. 26. Slanted floor for egg roll out.

Fig. 23. Cage front and sliding door.

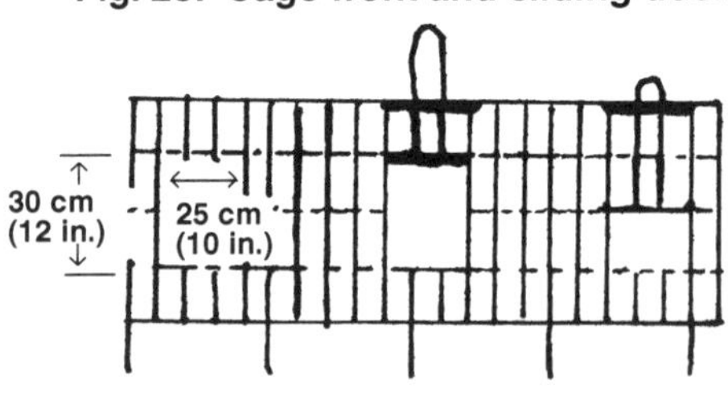

Fig. 27A. Four cages hung staggered from ceiling.

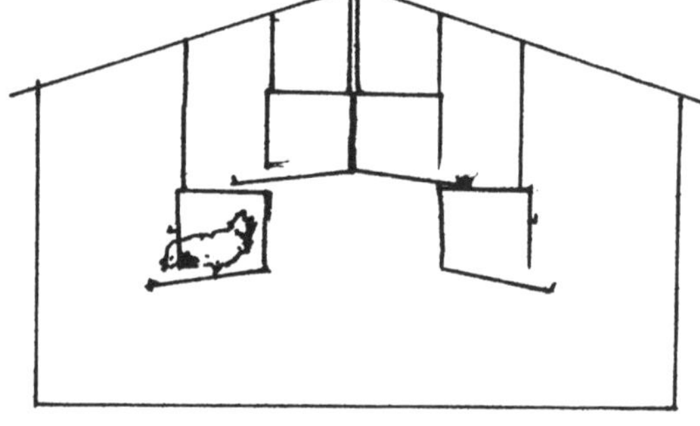

Fig. 27B

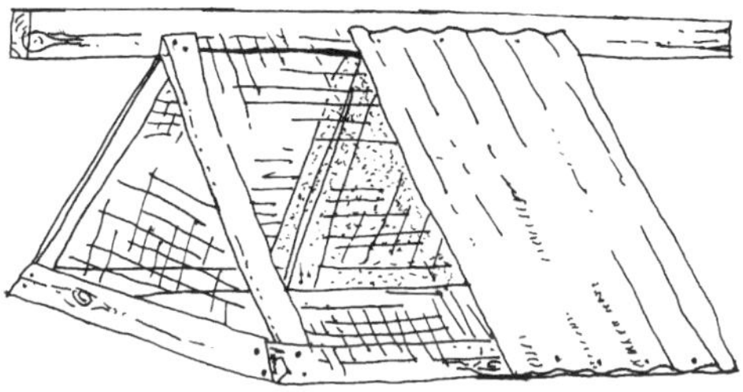

Fig. 27C

A simple method of keeping laying hens has been developed by R. G. MacGregor and L. Abrams, Department of Animal Health and Production, Faculty of Veterinary Science, Medical University of Southern Africa, P.O. Medunsa, 0204, Republic of South Africa.[1]

The aim of the program is to improve family nutrition through the provision of high quality protein. The program involves twelve young laying hens (pullets) being accommodated in cages which are suspended off the ground. Four eggs per day are sold to cover the daily feed cost, leaving the balance for daily household consumption. The hens are kept for one laying season only (one year) after which they are sold for human consumption at a similar price to which they were purchased. The money is used to purchase a new batch of hens (pullets).

The manure generated by the hens also serves as an excellent source of fertilizer for vegetables, contributing towards a balanced diet and enhancing family nutrition. The manure should be removed weekly to prevent fly breeding.

The plastic water bottles are soda or pop containers that have a dew drop nipple adapted to the lid. The plastic coated nipples used in commercial poultry operations work best.

Another aspect of the project has involved the employment of a physically disabled person to build the cages using locally available weldmesh. Communities in various parts of the country have also been involved in assembling their own cages which are sent to them in sections.

---

[1]Paper presented at the XX World's Poultry Congress, 2–8 September, 1996, India.

Section 5

Management

A. Incubating and Hatching. New poultry producers must choose between natural incubation of fertile eggs with broody hens or using an incubator. Natural incubation under broody hens or substitute turkeys, ducks, capons or geese requires little care from the poultry producer. However, it severely limits flock size and growth, and presents the possibility of disease transmission which may be controlled in artificial incubation programs. Most commercial poultry producers have come to rely upon a hatchery industry which has developed methods of artificially hatching large numbers of eggs.

Home-constructed incubators are still used for small flocks in some areas where the commercial hatchery business is not well developed. If a constant supply of electricity or kerosene is available, inexpensive incubators can be purchased or constructed which will hatch small numbers of eggs (Fig. 28). For chicken eggs the incubator must contain a thermostat (Fig. 29) which can be adjusted to maintain an optimal temperature of 37.5°C (99.5°F) which is the approximate temperature of naturally incubated eggs. Relative humidity should be maintained between 50 and 65% by use of water pans and adjusting the air flow. Chicks begin to hatch at 21 days and are often stronger if left unassisted. Although a few stragglers may live if given assistance, they are often too weak to survive. A small still-air, 50-egg capacity electric incubator can be built using a heat lamp with temperature regulated by a gas-filled thermostat. Eggs should be "turned back and forth" about 4 times a day during the early weeks of incubation. Mark each egg with an "x" to judge a turn of 45° each time but do not rotate.

Fig. 28. A homemade incubator.

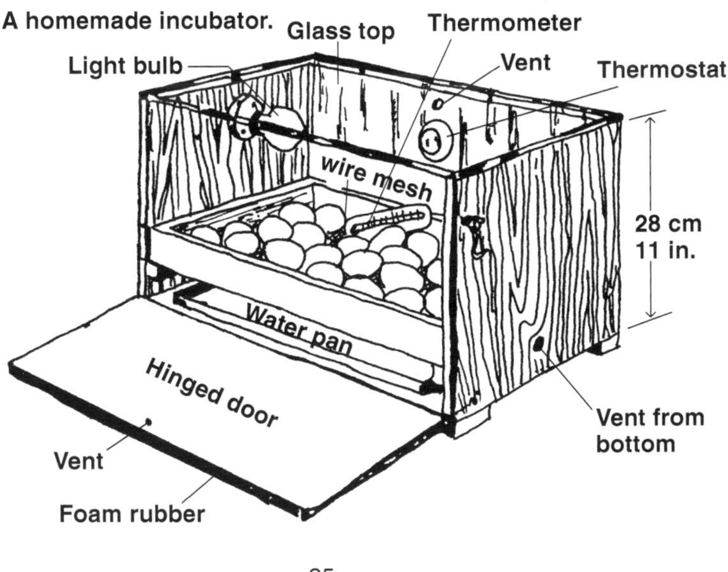

Fig. 29. Gas filled thermostat.

Fig. 30. Egg candler.

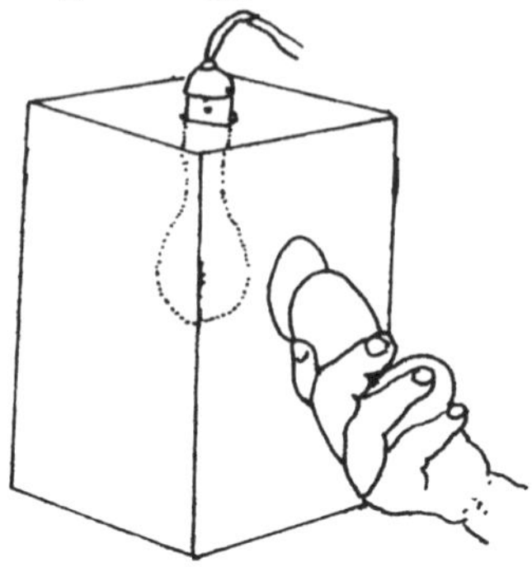

Fig. 31. Size of air cell after 7, 14, 18 days of incubation.

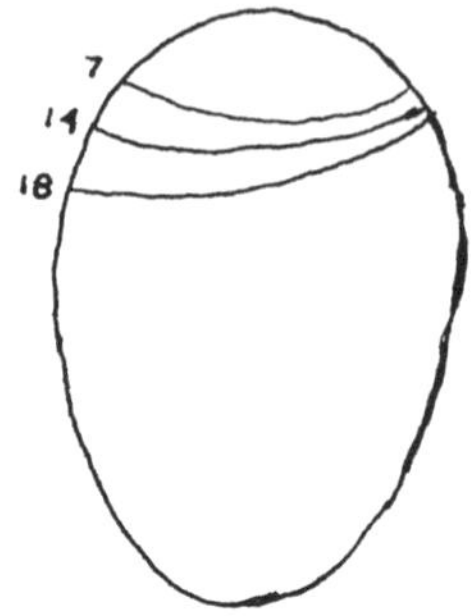

Since mechanical failures with home operated equipment are frequent, most poultry producers have come to rely on commercially operated *hatcheries* which employ mammoth forced draft incubators. These are designed to mechanically turn the eggs several times a day. Temperature should be maintained between 37.5 and 37.6 degrees with 50 to 60% relative humidity. For the nineteenth and twentieth days, the temperature should be dropped to 36.1 to 37.2 degrees with an increase in relative humidity to 75% just before the highest rate of hatching is reached. These recommendations depend on the type of incubator and size of the eggs being hatched, so pay careful attention to manufacturers recommendations. A back-up alarm system may automatically start a standby electric generator in case of power failure. Each incubator may well be provided with filtered air to prevent bacterial contamination from other incubators within the hatchery. Eggs are transferred from the incubator to a hatching unit on the 18th day of incubation.

A good commercial hatchery will supervise disease control on the breeder farm to prevent introducing diseases into the hatchery. Hatchery-transmitted diseases have been costly in the past. Thus the older custom of accepting hatching eggs furnished by individual breeders sometimes known as “custom hatching”, is rare. A thorough washing and disinfecting of equipment is essential for sanitary clean-up after each hatch. Services for vaccination, beak trimming and sexing are often provided along with supervision of the breeder farms to prevent egg-transmitted diseases such as *Pullorum* and *Mycoplasma.* Most of the newly hatched chicks are sold and delivered directly to the customer at one day of age. Some hatcheries brood and sell started chicks or pullets.

B. Candling Eggs. Progress in the developing embryo during incubation can be observed by rotating an egg in a beam of light. A simple candler can be made by putting a 3 cm (1-1/8 in.) hole to a can or box holding a 15-50 watt light bulb (Fig. 30). A flash light or narrow beam of sunlight in a dark room may also be used. Normal shrinkage of the air cell during incubation is shown (Fig. 31). If the airsac is more than 1 1/4 cm (1/2 in.) deep, the egg is probably spoiled.

Infertile eggs (white shells only) can be detected after 4-5 days of incubation. Two or three days longer may be required for tinted or brown eggs. If thin blood vessels may be seen as emerging from a dark red spot, the egg is fertile and the chick developing. If yolks and whites are clear the eggs remain wholesome, although laws prohibit their sale in some countries. Candling is also possible at 18 days while eggs are being transferred to the hatcher. Infertile eggs remain clear, contaminated eggs are detected by a darkened embryo inside and the absence of an air cell.

In some countries 12 to 14 day old incubated chicken eggs are used for "Baluts or Embryo Eggs" and they require refrigeration.

C. Brooding. If baby chicks are artificially hatched, the producer must assume the task of the mother hen in a role known as *brooding.* The attendant takes on serious responsibility for hourly attention during the critical early stages in the life of a flock of chicks. Constant inspection during the first two days may determine the difference between success and failure of a flock. In many countries women demonstrating maternal instincts do the best job of brooding during the critical first weeks. Three necessities must be constantly attended to: *heat, feed* and *water. Heat* for a brooder may be provided by electricity (Fig. 32-33), gas (Fig. 34), kerosene (Fig. 37), charcoal (Fig. 35) wood, coal, straw, or solar heat. Preparations include testing and adjusting the heat source 24 hours before the chicks arrive. The old rule of thumb for maximum brooder temperature has been 35°C (95°F) for the first week, and a decrease of 2.8°C (5°F) for each succeeding week until 21°C (70°F) is reached. Lethargic breeds of chicks require more heat than genetically active breeds. Thus some authorities suggest a starting temperature of 30-35°C, varying with breeds. A range of temperatures in the brooding area gives chicks a choice in selecting an optimum temperature.

An inverted box-type brooder with a protected light bulb hanging 20 cm (8 in.) from the floor furnishes a satisfactory brooder for about 50 chicks (Fig. 32). A confined area outside will permit chicks to run outside through the cloth curtain. An infra-red light bulb adjusted to hang 45 cm (18 in.) from the floor is a good heat source for small numbers of chicks (Fig. 33). A hood, which is usually constructed of sheet metal, may be hung or suspended by small posts or bricks with the edge 10-12 cm

Types of Brooders

Fig. 32. Light bulb brooder.

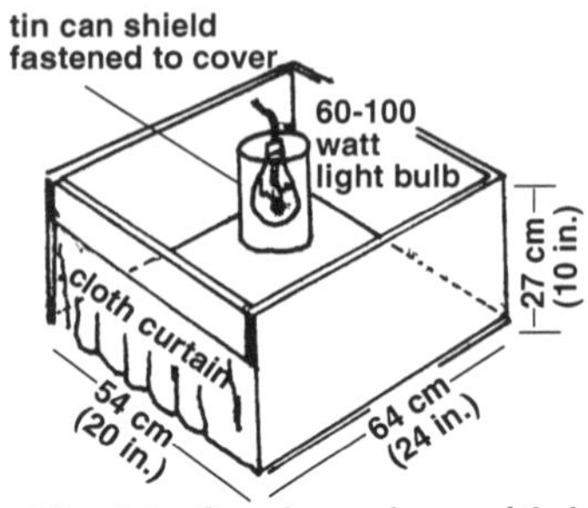

Fig. 33. Infra-red heat lamp.

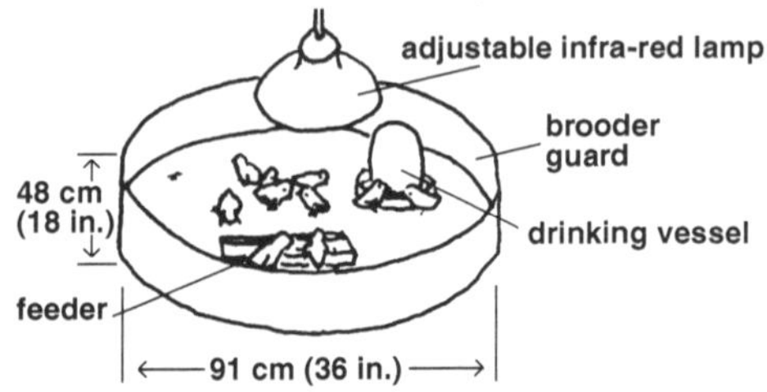

Fig. 34. Gas brooders with hood.

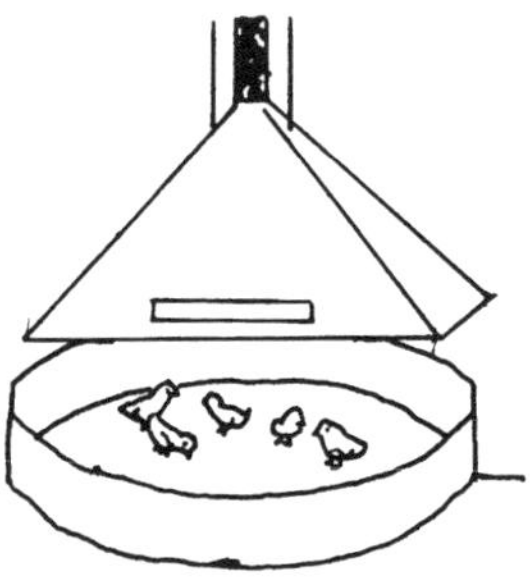

Fig. 35. Oil barrel—charcoal.

Fig. 36. Basket brooder.

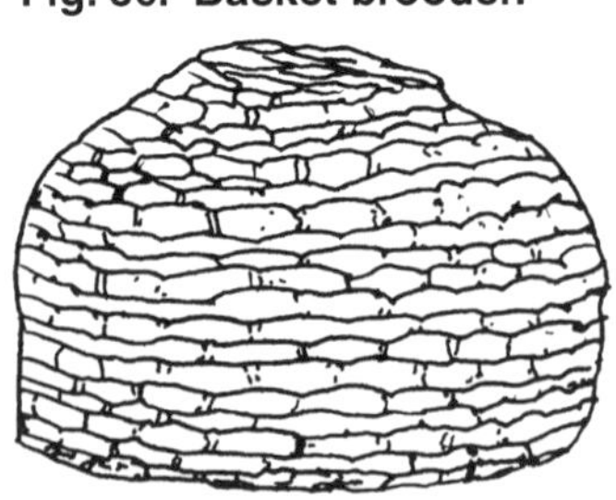

Fig. 37. Kerosene brooder.

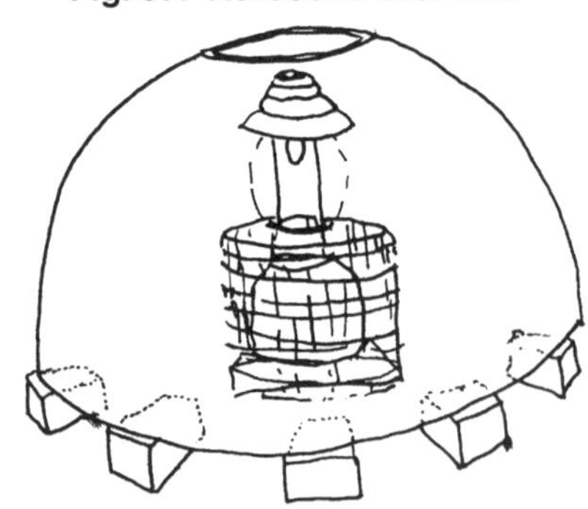

(3-4 in.) above the floor. Basket brooders may be constructed with a kerosene lamp propped up on bricks as a heat source. Caution to prevent fires is required with any heat source.

For the first week a ring known as a "chick guard" (it may be cardboard, metal, bricks, wood or any material that is clean and will keep the chicks close to heat source) or "brooder guard" is often placed beyond the rim of the brooder to prevent chicks from wandering too far from the heat source (Figs. 33-35). An experienced observer will be guided in temperature regulation by the chick behavior (Figs. 38-41). Less experienced care-takers should frequently record the temperatures indicated by a thermometer located at the level of the chicks' backs. If chicks are huddled together and making the chirping sounds of discomfort, more heat may be necessary. If chicks are distributed in a ring as far possible from the heat source, less heat should be furnished (Fig. 40). Constant attention for young chicks is especially important because outside temperature changes alter the amount of supplemental heat necessary.

Brooder Management

Fig. 38.
Chicks too cold. Lower lamp.

Fig. 39
All is well; conditions just right.

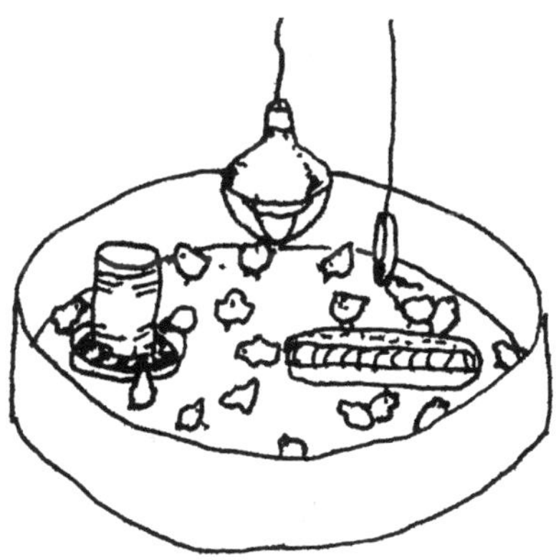

Fig. 40.
Chicks too hot. Raise lamp.

Fig. 41
Cold draft. Plug it up.

Fig. 42. Feed and water placement within brooder guard. Arrangement of 4 waterers and box lids (temporary) to be replaced by feeders.

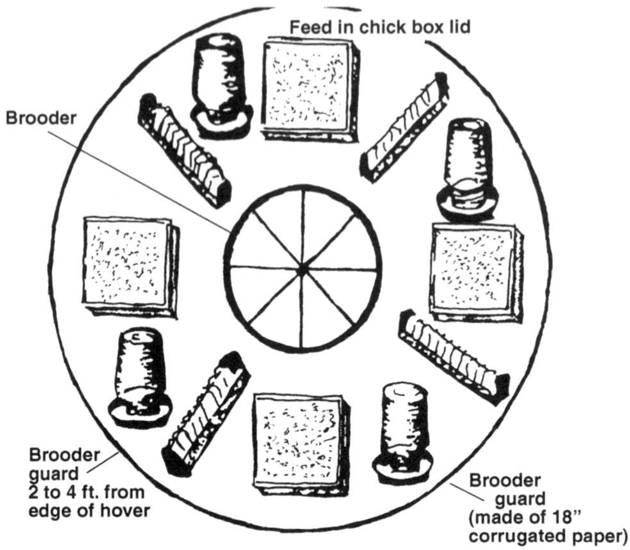

Feed must be presented in a manner that will attract the curious chick during its first day of brooding. Mixed mash, or crumbles spread out on newspapers or an inverted chick box top is frequently used. Four feeder and 4 watering stations are often placed alternately just inside the brooder ring (Fig. 42). Place baby chicks right on top of the feed. During the first week temporary box-top feeding stations are gradually replaced with 4 small chick feeders. Each 100 chicks requires 2.5 m (199 in.) of feeding space.

Water must be found by each chick during the first day even if some must be picked up by hand and their beaks inserted in the water. During the first 2 weeks no chick should be more than 1 m (3 ft.) from a waterer and thereafter not more than 3.5 m (10 ft). Daily requirement of water for 100 chicks approximates 2 liters during the first week. Size of the waterers will need to be increased as the chickens grow (Figs. 5-12). One hundred hens in lay may require 36 liters or more per day. Spillage on to the litter needs to be guarded against (Figs. 10-11).

D. Rearing. Use of added heat in brooders may be discontinued after 3 weeks, or earlier in warm climates. The producer has a choice of different systems of management. The flock may be reared on *free range* (Fig. 1), *limited range* (Fig. 2), in *pens,* in *cages* or in *confinement within houses* (Fig. 3 and 4).

Advantages of *range* rearing include 1) reduced feed costs as birds secure much of their own diets by scavenging for green fodder, insects, and by scratching in manure from large animals, and 2) reduced housing costs. Disadvantages include 1) loss from predators, and 2) it requires close flock shepherding. A *limited range* requires less flock supervision. With natural brooding the hen will care for the chicks. They should be in a protected place during the day and at night. Predators such as dogs and hawks can be a problem any time.

Advantages of *pen* rearing include more protection from weather and predators. Disadvantages include the cost and responsibility for furnishing all essential nutrients in the feed and cost of complete feeding, fencing, and problems arising from wet pens in rainy weather which may result in disease and parasite problems.

Cages save space and if suspended to avoid contact with droppings, birds may be protected against some diseases and tick parasites. Cages have a higher initial cost and require a convenient manure disposal system to prevent flies and odors.

Confinement rearing permits greater protection against predators and may prevent disease if sanitary measures exclude visitors, older chickens, pets, or fomites. All essential nutrients must be furnished in the feed. Recommended space per bird should be 650 sq. cm (100 sq. in.) for the first 5 weeks, and increased to 2,750 sq. cm (3 sq. ft) as it reaches maturity. The floor is usually covered with 5 to 10 cm (2-4 inches) of *litter.* Type of litter selected depends upon local availability and cost. Preferred materials include: peat moss, chopped oat, rice, or wheat straw, softwood or hardwood chips, sawdust (only after 4 weeks), shredded corn stalks, ground corn cobs (some danger from molds), broadleaf leaves, rice, peanut or coffee bean hulls, shredded sugar cane stalks, newspapers, and sand (last choice). Poor quality litter requires somewhat greater floor space to stay dry. If litter becomes wet due to leakage of waterers or rain, wet material should be replaced. Fine grass, straw or hard particles of incompletely ground feedstuffs may cause a blockage known as "crop-bound" (Fig. 43). If such chickens are otherwise healthy they should be sacrificed for meat since there is no practical remedy to relieve this condition.

Fig. 43. Crop-bound chicken.

Fig. 44. Killing a chicken for a post mortem examination.

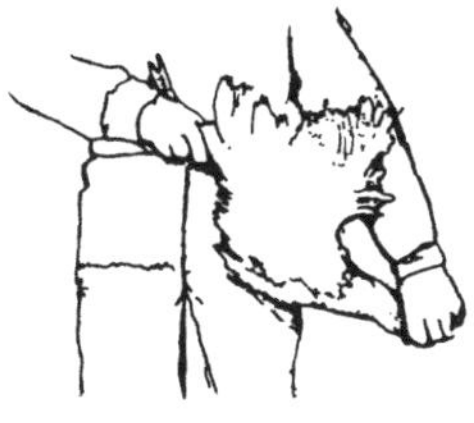

Fig. 45. Metal cone for weighing.

Fig. 46. Zones for beak trimming.

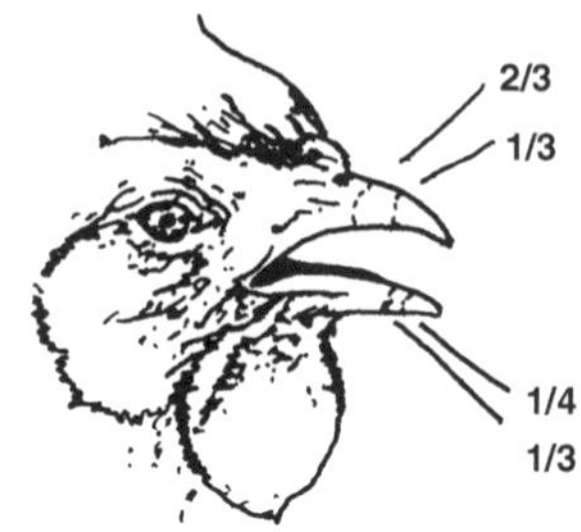

Fig. 47. A catching hook.

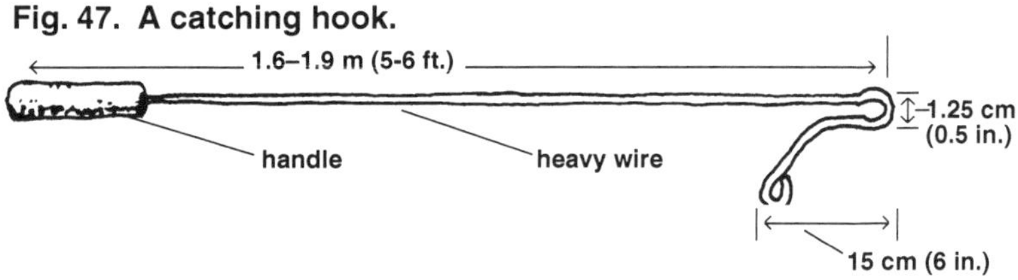

Cannibalism. Beak trimming is a method of cutting or burning off the tip of the upper and lower beaks (Fig. 46). If properly done this minor operation prevents cannibalism which is a bad habit of pecking at feathers or injured areas of other birds. Some broiler and layer producers order beak trimming of all birds at the hatchery. For broilers, only about 1/3rd of the top beak is removed. For layers which will be alive longer, 1/3rd to 2/3rds of the upper and 1/4th to 1/3rd of the lower beak is removed. Scissors or a sharp knife may be used if a debeaking machine is not available. Dip the freshly cut beak into feed if bleeding occurs. With birds on the range equal amounts of each beak should be removed only after the onset of cannibalism. Cutting too deep into the nares should be avoided. Where electricity is available a cauterizing machine may be used. One such beak trimming machine is sold by the LYON ELECTRIC CO., 2765A MAIN STREET, CHULA VISTA, CA 91911 (USA) FAX 619/420-1426.

Other methods of preventing cannibalism are sometimes required. At the first sign of feather pulling or blood the victims should be removed from the pen. Raising the birds in very dim light, a practice known as "dark house brooding" is an effective means of preventing cannibalism. This requires a windowless house, fan ventilation, and dim light furnished by a few small light bulbs. At least .5 foot candles (5.4 lux) of light must be provided. Care is critical in placement of feeders and waterers to prevent birds from becoming dehydrated and losing weight in the dark.

E. Layers. Debates continue over which is the best management system for layers. Free-range provides access to sunlight (which prevents rickets) and gives birds access to green fodder. Eggs from such sources have sometimes sold at a premium because of increased pigmentation in egg yolks. However, general production costs are considerably higher than eggs produced with confinement rearing. Scavenging for feed on range may save some costs for feed. Partial restriction by fencing provides increased protection but may also increase exposure to diseases and parasites if pens get wet.

Confinement rearing on litter in floor pens, on elevated slatted floors (Fig. 52), or in cages is now widely practiced. This requires a feed source which is nutritionally fully balanced. Capital investment for housing is lower with litter on floor pens than cages. Cages conserve space, facilitate egg collection, save on labor costs but are expensive to install. Manure accumulations under cages or raised slatted floors sometimes creates housefly problems as they breed in damp manure.

Additional *lighting* applied at the proper time in the life cycle of hens may greatly increase egg production. With leghorn types the optimum time to increase lighting exposure is usually at 20 weeks of age. With dual-purpose breeds 21 weeks and broiler breeders 22 weeks is often the best time. Instructions in management guides provided by breeding organizations should be closely followed. These instruct on weight of the pullet (Fig. 45) at the time increased light is to be applied. Add one hour of light each week until 15 hours total is reached. Egg size may remain small if increased lighting occurs too early. Switching lights on and off during the night is another management program sometimes used to stimulate feed consumption and rate of growth of broilers. In large commercial operations light switches are controlled by time clocks.

Another factor in numbers of eggs produced is the regular daily collection. If a yard hen is permitted to keep her eggs in the nest, her production may be limited to 30 eggs per year in two clutches of 12-15 each. If collected daily she may lay 120 eggs. If 15 hours of daylight are provided she may lay 250 eggs.

Community projects have been developed for Household Egg Production.[1] The program allows a family to have 10 or 12 ready to lay females. They are kept in a suspended or hanging wire cage that is divided so 2 or 3 chickens are kept in each compartment. The manure is cleaned often, to reduce fly breeding, and used on the garden. After one year of lay the hens are sold, the cages are cleaned and new ready to lay hens are obtained.

Nests. With floor-pen management nests should be provided for layers before they begin to lay. They may be constructed of wood (Fig. 48), sheet metal, wire mesh, clay, mud bricks, or woven mats (Fig. 50). Individual nests should provide a space of about 0.12 cubic m (1 cubic ft.). They may be placed on the floor (Fig. 48), on stilts (Fig. 49) or against the wall and in the darkest part of the house (Fig. 51). They should be lined with fresh litter and kept clean to prevent production of dirty eggs. If tiered or hung above the floor, nests should be provided with a perch in front. Colony nests, which may be used by several birds at one time, may be satisfactory if they are four times the size of a single nest. With large slatted floor houses, nests are arranged for the convenience of the caretaker (Fig. 52). Allow 25 nests for 100 hens and close them at night to prevent roosting and open at daybreak. This will keep the nests clean and reduce dirty eggs.

Eggs should be collected frequently and stored in a cool place to prevent rapid deterioration in quality. Some producers make 3-5 collections a day in hot weather. Eggs held at 37°C (99°F) drop drastically in quality after 3 days while they may keep 100 days at 3°C (37°F). Three degrees may be best for long term storage, but 10°C (50°F) is fine for shorter periods. These guides are for commercial eggs only. Hatching eggs require special care and should not be kept more than about 10-14 days at 18°C (65°F).

Roosts. Since birds traditionally sleep in trees, they are comfortable if the producer provides roosts. However, roosts are unnecessary. Hens require about 20 cm (9 in.) of roost space per bird. Roosts do concentrate droppings which may be salvaged for fertilizer. If roosts are built, house design should be planned for ease in cleaning out droppings.

Culling is the practice of removing non-layers from a flock. Considerable savings in feed costs can be effected by periodically catching and removing unproductive birds from the flock. However, caution in selection is suggested since birds which have recently molted may recover

---

[1]Household Egg Production Program, compiled by Professor L. Abrams and Mr. G. MacGregor, Department of Animal Health Production, Faculty Veterinary Science, Medical University of Southern Africa, South Africa (012) 529-4751/40; E-mail: *gavin@mcd4330.medunsa.ac.za.*

their ability to lay after a period of rest. A "catching hook" (Fig. 47) facilitates their recovery. Signs of a non-layer are a shriveled, pale, scaly comb, yellow pigment in the eye ring and the beak, dry rough wattles, small, round, yellow (Fig. 54), puckered vent (Fig. 56) with space between the two pubic bones for only 1 finger (Fig. 58), and 2 fingers between the pubic and breast (keel) bones. Note that the reference to pigmentation applies only to genetically yellow skinned birds receiving yellow pigments in the feed. In contrast, active layers have large, red,

Nests

Fig. 48. Individual wooden nest on floor.

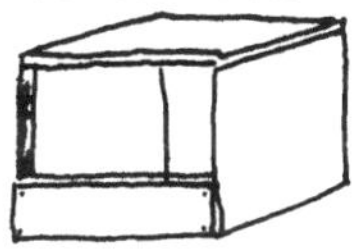

Fig. 49. Double nest of wood on stilts.

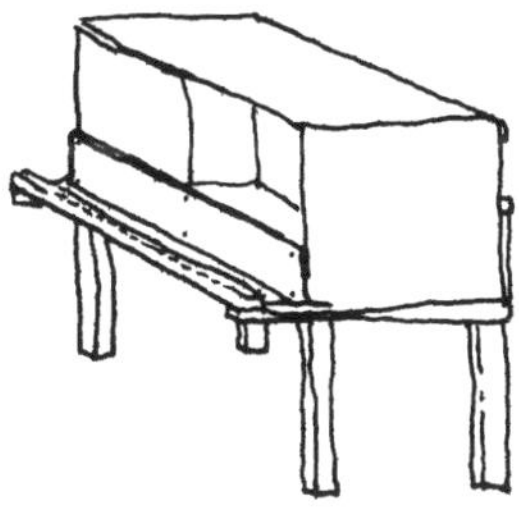

Fig. 50. Basket made of reeds and clay.

Fig. 51. Nests hung against the back wall. Perch is hinged for night closure.

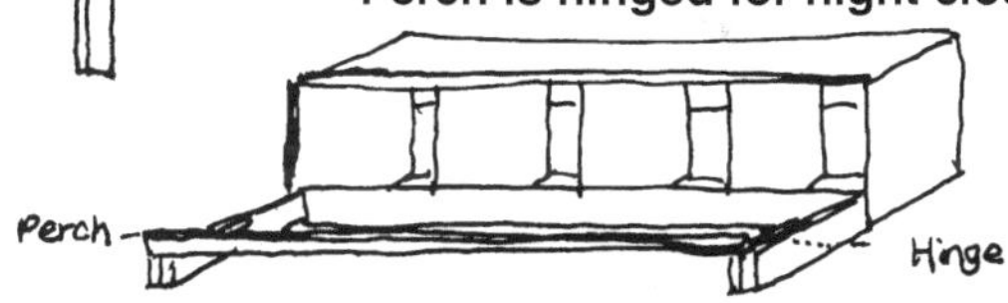

Fig. 52. Slat-litter house with convenient egg collection.

Layers vs. Non-layers

Fig. 53. Hen in lay.

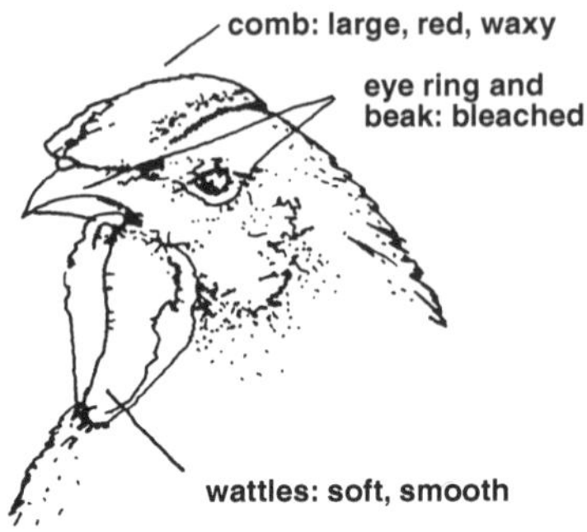

Fig. 54. Non-layer.

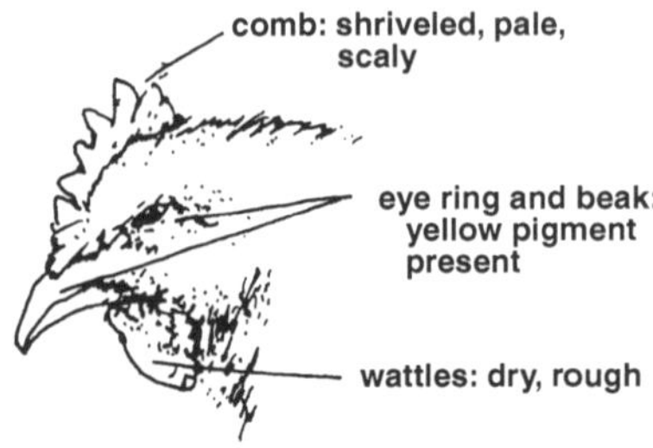

Fig. 55. Hen in lay.

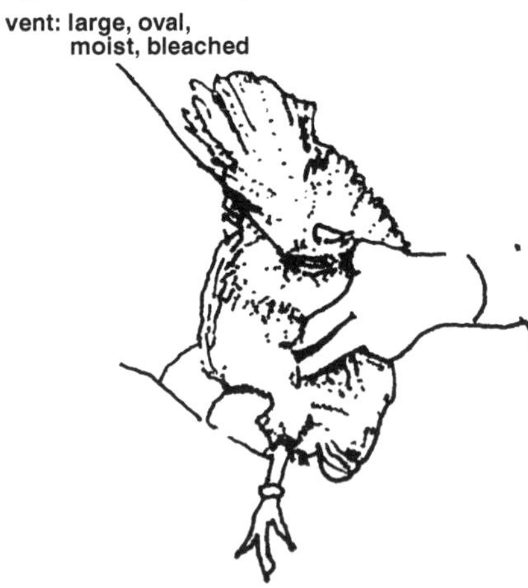

Fig. 56. Non-layer.

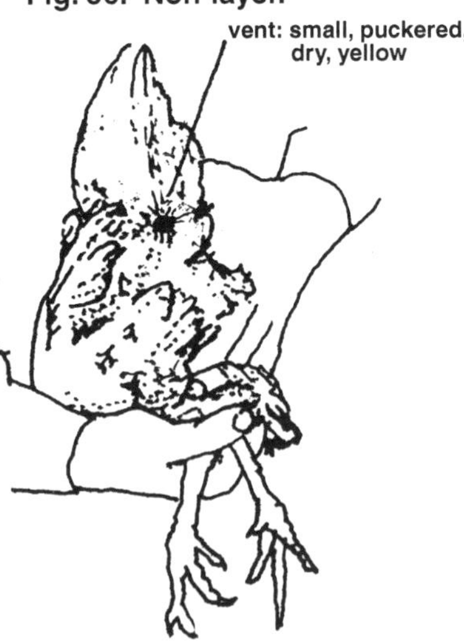

Fig. 57. Hen in lay: 3 fingers.

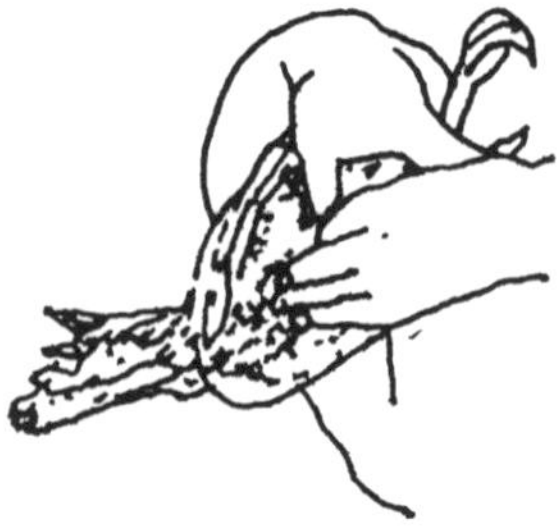

Fig. 58. Non-layers: 1 finger.

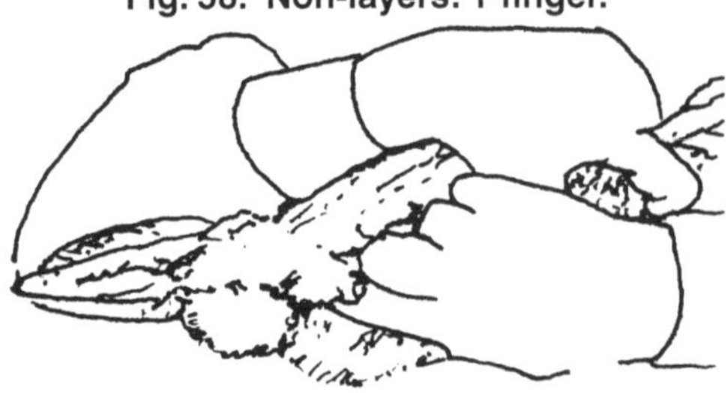

waxy combs, bleached eye ring and beak, soft and smooth wattles (Fig. 53), large, oval, moist, bleached vent (Fig. 55), with room for 3 fingers (Fig. 57) between pubic and 5 fingers between pubic and breast bones. Since even good layers may sometimes take 2-3 week pauses between clutches, caution is necessary to prevent the accidental selection of some potential good layers.

F. Breeders. Each of the commercial breeding establishments selling male or female parent stock provides a manual with suggestions for rearing their birds to obtain optimum production of meat or eggs in the offspring. To judge maturity rate, birds are selected at frequent intervals to be weighed as flock samples (Fig. 45). Placing the bird in a cone-shaped metal container head first is a convenient method. Different directions are sometimes given for feeding and rearing male and female lines. Since the price of these breeders as baby chicks may be several times the cost of other chicks, all management suggestions are usually closely followed.

G. Records. Daily records should be made on charts conveniently located within the poultry house. Ruled lines with the calendar date on a card or paper attached to a clipboard are suggested. These should include space to record mortality, date of delivery of chicks, feed delivery date and source, feed consumption, daily egg production and records of any management or disease problems. Although this chore is distasteful to many poultry producers, these records become invaluable in planning for future poultry projects and disease control programs.

Section 6

Nutrition

Energy

Proteins

Minerals

A. Introduction to Poultry Nutrition and Feeding

If poultry is permitted to forage in open fields or barnyards, birds usually manage to obtain enough of a balanced diet to survive. A household flock of 10 to 20 birds may require no feed in addition to table scraps, green fodder and insects from nearby gardens. Slightly larger flocks survive better if fed a few handfuls of corn or other grain. Encouraged by the income from the sale of these free-range flocks, the planner may be tempted to greatly expand flock size, sometimes forgetting that this may require the purchase of most of the feed. Plans must be made to provide a balanced diet with prospects for enough financial return to justify the added expense. Only expand the flock if a market is available for the eggs.

The producer may be well-advised to *start small* and increase flock size gradually after some experience with feeding and marketing poultry. Commercial producers calculate that 50 to 70 percent of the expense of producing chickens is due to the cost of the feedstuffs. Thus advance planning for a reliable supply of feed is necessary. As birds grow, their feed requirements increase each day. "Full feeding", a practice of keeping feed in front of the birds at all times, is often recommended. Estimates should be made on the total feed requirements to rear the flock to market age. Weight of feed required will be at least 2 or more times the expected market weight of fast-growing meat birds. A feed conversion ratio (feed weight divided by market weight of the birds) of 2 is considered good for broilers. To obtain such a ratio requires 1) good broiler breeder stock 2) a nutritious and well-balanced diet 3) good management and 4) disease control. The sale value of the bird minus the cost of the feed may be more important than the feed conversion ratio. Less expensive feed sometimes leads to slower growth but a greater profit.

Many poultry producers have lost all profits due to flock starvation when essential feedstuffs have unexpectedly disappeared from the market. Other losses, which may go undetected by the producer, occur due to nutritional deficiency diseases. Since grain products constitute the bulk of many mixed poultry feeds, competition with human food supplies becomes a major consideration. The enterprising poultry producer usually expends considerable effort in locating inexpensive feed ingredients which are unsuitable for human consumption.

Although the backyard producer has come to think of grain as the ideal supplement for scavenging chickens, a variety of feed ingredients is essential if chickens are to be raised in confinement. To insure rapid growth and egg production most commercial producers have come to rely upon a nutritionally balanced mixed feed given as mash, crumbles, or pellets.

B. Ingredient Sources to Meet Nutrient Needs:

For optimal performance some 40 different chemical substances, or nutrients, have been identified as essential in the diet. A brief summary of six classes of ingredients is presented here:

Feedstuffs Rich in Energy. Hunger is the driving force which stimulates a chicken to eat and satisfy its *energy requirement.* Energy is required for muscular activity to carry on chemical processes like digestion and for producing heat. The feed ingredients must be proportionately balanced with energy-containing foods to supply all nutrients in proper ratios.

Feedstuffs providing energy are often subdivided into three categories: *carbohydrates, fats* (lipids) and *proteins.* Although surplus proteins do furnish energy, their need to satisfy growth requirements plus their added cost makes them too expensive to consider for energy alone. Carbohydrate sources may be subdivided into *starches, sugars* and *cellulose (a fibrous plant tissue).* Starches and sugars usually furnish a major portion of the energy requirements. These are usually supplied by feeding grains. Although celluloses (indigestible fibers) may be a major constituent of the feed, they are not utilized by poultry. Fats or oils provide an added energy reserve. If measured by weight they produce more than twice the energy yielded by carbohydrates or proteins. The energy value of feeds is usually expressed in terms of kcal or Joules per unit of feed. Both are metric system units for expressing energy, kcal are specific measures of heat energy. One cal = 4.184 Joules = heat required to raise 1.0 gram of water one degree C. Fats yield about 9 kcal per gram, while carbohydrates and proteins yield about 4 kcal of metabolizable energy for each gram.

Feedstuffs Which Contribute to Energy Needs

Corn (Maize) This cereal is the preferred energy source in many poultry diets and often constitutes as much as 60 to 70% of the ration. Corn based diets must be supplemented for deficiencies with certain amino acids, vitamins and minerals. Yellow corn is preferred over white if yellow pigmentation of egg yolks or skin color is preferred by consumers.

Wheat Although frequently more expensive than corn, wheat may be used up to 70% of the diet. By-products of flour manufacturing such as wheat middlings and bran may also be used. Bran contains considerable fiber which cannot be used by poultry, but is a source of some vitamins.

Other grains are also supplied as energy sources. Each one has somewhat different nutritional properties. *Sorghums (milo, millet, kafir corn), barley, and oats* are frequently used. Crushing or grinding is usually recommended and in some feedstuffs husks should be removed if possible.

Rice screenings are sometimes included in poultry rations after lighter husks have been removed. Such products must be fed while fresh after winnowing.

Cassava (manioc) roots contain large quantities of starch and may be incorporated in levels up to 25% of mixed feeds after drying to remove toxic hydrocyanic acid.

Potatoes and yams (sweet potatoes) have been used as a short-term energy substitute for grain products.

Molasses may be used in levels up to 5% as a carbohydrate source.

Cane juice has been used in broiler diets. Cane juice and molasses are valuable sources of energy, but they contain almost no protein.

Fats—oils and tallow from plant or animal sources are sometimes added to mixed feeds to supplement energy supplies. Between 3 and 6% of the ration is generally recommended.

The bird's energy intake needs to be in balance with the protein and other nutrient intakes for the optimum utilization of the diet for meat and egg production.

Feedstuffs Rich in Proteins and Amino Acids. All new body cells of a baby chick are synthesizing distinctive new proteins from as many as 20 different amino acids. Amino acids, the building blocks of proteins, are derived from plant and animal feedstuffs. Some amino acids cannot be synthesized by chickens (arginine, lysine, histidine, leucine, isoleucine, valine, methionine, phenylalanine, tryptophan, and threonine). They are called essential, or indispensable, amino acids and must be present in each poultry ration. Other amino acids can be synthesized only in some situations (cystine, glycine, serine and tyrosine).

General signs of amino acid or protein deficiencies include growth depression, abnormal feather development, decline in egg production, and especially reduced egg size in laying hens. Deficiency of the amino acid lysine causes lack of pigment deposit in black or reddish colored feathers. One-third of the protein in a hen's diets may come along with energy from the grain source. The key to protein feeding is to compliment the chosen grain source. Most grains are low in lysine and high in methionine, so the best protein supplements are likely to be legume and/or animal protein meals, which are high in lysine and low in methionine.

Feedstuffs Which Contribute to Protein Needs

Plant Sources. *Soybeans and Soybean oil meal.* Soybean protein is rich in lysine, and therefore an excellent compliment to grain proteins. Soybeans must be cooked or heat treated to destroy growth-inhibiting substances. Heat processed whole beans make an excellent high protein (36%), high energy, supplement. Meals with the oil removed are excellent protein supplements containing approximately 44-49% protein.

Peanut Meal (Groundnut) may contain 47% protein but is deficient in some amino acids, particularly lysine, methionine and cystine.

Cottonseed Meal may contain up to 43% protein but a toxic compound ("gossypol") must be removed by special treatment before it is fed. If the meal is not treated growth depression occurs in chicks and egg yolks are discolored.

Meals *from sunflower, safflower, sesame, tung nut, kapok oil, palm kernel, rubber seed, copra, field peas, navy, lima, winged, and other beans* together with *brewers and distillers by-products* have been used for protein supplementation. Each product can be fed at low levels, but all have limitations which need to be considered in feed formulation. The amounts of these meals that may be economically fed are dependent on the source of each ingredient. The best approach is to buy from a reliable source, and learn from trial and error how much of each ingredient can be fed. If the source and quality of an ingredient changes, the birds response may also change.

Animal Sources. *Meals made from fish, meat, blood, offal (visceral wastes), and feathers,* contain varying levels of essential amino acids and minerals. Phosphate supplements are often expensive and animal protein meals are usually good sources of available phosphorous. Surplus *dairy products* such as *milk* and *whey* are also good sources of scarce amino acids and some essential minerals. Buttermilk is an excellent supplement to grain-based diets; it increases palatability and adds nutrients. It may be poured on the feed in the trough. To prevent mold formation, add only enough for a few hours consumption in hot climates, and not more than will be consumed in one day, even in cool climates.

Synthetic Amino Acids. Two amino acids, *lysine* and *methionine,* that often limit the nutritional quality of feedstuffs, are now available at competitive prices as feed supplements from commercial manufacturing processes. Synthetic methionine hydroxy analog is also available as an economic methionine source for poultry. This analog is quickly converted to methionine in the bird's body, taking advantage of waste nitrogen available in the bird.

Other Ingredients

Many other ingredients may be included in poultry diets if available at reasonable prices. These include grain crops and animal and plant by-products. Each contains different amounts of protein, energy, vitamins and minerals. Examples: *bananas, carob pods, coconut products, copra meal, cowpeas, dried dates, gram, horse beans, alfalfa (lucerne) meal, lupins, lobster waste, manioc, millet, molasses, palm kernel cake, pineapple leaf meal, rice bran, rice paddy, simsim (sesame) cake, snails, sweet potatoes, dried worm meals, and yucca products.*

3. Vitamin Supplements. Some 13 different vitamins are required to produce healthy chickens. Deficiency symptoms (Table 6.1) sometimes indicate the absence of one specific vitamin, but more frequently, these symptoms indicate shortages of several vitamins and/or minerals.

TABLE 6.1 Signs of Vitamin Deficiencies and Natural Sources

Vitamin	*Signs of Deficiency*	*Rich Sources*
Vitamin A retinol	"nutritional roup", ataxia, eye infections, urate deposits, reduced egg production poor hatchability	green forage corn gluten meal yellow corn (carotene) fish oil
Vitamin D	rickets, soft bones, poor growth, feather- reduced egg production	chicks in sunlight synthesize this vitamin
Vitamin E tocopherol	"crazy chick disease" encephalomalacia, edema, muscular dystrophy poor hatchability	alfalfa meal, vegetable oils wheat germ
Vitamin K	hemorrhaging, poor blood clotting	alfalfa green pasture fish meal
Vitamin B_1 thiamin	"star gazing" loss of appetite, weight loss, convulsions laying stops	leafy feedstuffs milk and grain products oil seed meals
Vitamin B_2 riboflavin	curled toe paralysis, reduced growth and hatchability	alfalfa, green grass milk by-products fermentation products
Pantothenic acid	dermatitis of feet, poor growth and feathering, lesions of mouth, eye lids	alfalfa, green grasses, milk products distillers' fermentation products

TABLE 6.1 (continued)

Vitamin	*Signs of Deficiency*	*Rich Sources*
Nicotinic acid	slipped tendon (perosis) hock enlargement	wheat middlings corn gluten meal
niacin	poor feathering	alfalfa
Vitamin B_6	dermatitis poor growth and coordination, convulsions meat and fish meals	milk products soybean meal
Biotin	poor growth, dermatitis in feet, mouth, eyes, slipped tendon, reduced hatchability	green pasture soybean meal, grains, dried yeast milk products
Folacin folic acid	poor growth slipped tendon anemia, poor feathering and hatchability	leafy vegetables
Choline	slipped tendon="perosis" poor growth and egg production	fish and milk products soybean meal
Vitamin B_{12}	anemia, poor growth and hatchability	fish and animal by-products animal manures

Many of the vitamins are readily available in grains, green fodder or natural fermentation products if scavenger management is used. The B complex vitamins appear in yeasts and fermented animal manures. Supplemental B complex vitamins and vitamin A are sometimes furnished in small poultry operations by providing cut-up green fodder. Although vitamins from this source deteriorate rapidly, they may be preserved by drying. For this reason alfalfa meal is often used as a feed supplement. The poultry producer should be on the lookout for milk by-products such as whey and brewery waste products when vitamin supplements are needed.

The alternative to natural feed sources for vitamins is the inclusion of a concentrate as a feed supplement. Most of the vitamins are inexpensively produced from chemical processes or from fermentation waste products. Concentrates of vitamins as well as trace mineral additives are available commercially. Shipping charges are low since only small quantities are required. For example a typical premix added at the rate of 0.05% of the diet provides (per kg/diet): vitamin A, 5,500 IU; vit. D_3, 1100 ICU; vit. E, 11 IU; riboflavin, 4.4 mg; Ca pantothenate, 12 mg; nicotinic acid, 44 mg; choline Cl, 220 mg; vit. B_{12}, 6.6 mcg; vit. B_6, 2.2 mg; menadione, 1.1 mg; folic acid, 0.55 mg; d-biotin, 0.11 mg; thiamine, 2.2 mg (as thiamine mononitrate); and an antioxidant to increase the vitamin stability. Purchase of such a premix should be seriously considered by any poultry producer who must mix his/her own feed.

4. Mineral Supplements. Nine or more inorganic elements are often classified as essential minerals. They must be supplemented to the diet if absent from the major feedstuffs. Calcium, phosphorus, sodium, potassium and chloride must be present in relatively large quantities. Potassium is generally available from practical ingredients, while sodium and chloride are provided by salt. Calcium and phosphorous supplements need to be carefully balanced. Suggested feed levels for calcium are 0.9 to 1.0 % for pullet starter and broiler feeds, and 3% for layer feeds. Since large quantities of these minerals, with costly shipping charges, are required, a local supply of limestone and bone meal is desirable. Calcium and phosphorus are necessary for good bone structure and egg shells. An absence of either, or a vitamin D deficiency, causes a condition of soft bones called rickets. A balance of 2 parts of calcium to 1 part of *available* phosphorus should be maintained for growing chickens. Two-thirds of the phosphorous from plant materials is considered non-available. Layers need no more than 0.32% available phosphorus. Cooked or pasteurized hatchery waste containing egg shells has been used as a source of calcium. Crushed limestone or oyster shells are often provided in separate containers so hens may supplement their diet by free choice of calcium. This practice is unnecessary if layer feed formulation is properly balanced.

TABLE 6.2 Signs of Deficiencies in Inorganic Elements and Feed Sources

Inorganic Elements	*Deficiency Signs*	*Feed Sources*
calcium	rickets, soft bones thin egg shell growth retardation	oyster shell ground limestone (low or no fluorine) bone meal
phosphorus (phosphates)	rickets reduced egg production thin egg shells	cereals bone & fish meals defluorinated phosphates milk products

TABLE 6.2 (continued)

Inorganic Elements	*Deficiency Signs*	*Feed Sources*
sodium	poor growth cannibalism	common salt animal products
potassium	poor growth	plant feedstuffs
iron	anemia	plant and animal products
manganese	slipped tendon (perosis) "star-gazing"	rice hulls wheat middlings
selenium	severe anemia exudative diathesis poor egg production	fish meals brewers yeast
iodine	goiter	fish meal brewers yeast

Several other inorganic mineral elements are required in minute quantities (Table 6.2). Since these "trace" elements are light in weight they may be shipped inexpensively in trace-mineral premixes. Recommended levels have been established for iron, iodine, zinc, manganese, copper, magnesium and selenium. The trace mineral status of feedstuffs is highly dependent on the quantity present in the soils in which they were grown. When all the feed ingredients in a diet come from one geographic region, the chance of trace mineral deficiencies in animals is greatly increased. Other trace elements which may be important under special conditions include fluoride, nickel, tin, and vanadium.

Diseases caused by trace mineral deficiencies include: iron—anemia, iodine—goiter, and zinc—enlarged hock joints. The absence of dietary selenium causes a disease condition with appearance of gelatinous green pustules under the skin ("exudative diathesis"). This condition appears in limited areas of the world. In other areas selenium levels are high enough to prove toxic to large animals. Feedstuffs grown in soils containing low selenium levels require added traces at the level of 0.2 mg per kilogram of diet. A typical mineral premix used at the rate of 0.05% to the diet would add in mg per kilogram of diet: Manganese, 60; Zinc, 50; Iron, 30; Copper, 5; Iodine, 1.5; Selenium, 0.1. Wood ashes often provide many of the trace elements in scavenger poultry management.

Supplying insoluble grit in the form of small stones has sometimes been recommended for confinement rearing to assist the gizzard in grinding feedstuffs. However many feeding trials indicate that this practice is no longer considered necessary when feed ingredients are routinely ground. Scavengers or yard chicken will search out their own grit sources.

5. Water. Chickens require free access to clean, preferably cool drinking water at all times. All growth, body maintenance, activity, and egg production requires water. Water makes up 85% of the body weight of a chick. Although chicks can survive for several days or even weeks without feed, a 20% weight loss due to water deprivation results in death. Molting may be induced by short periods of water deprivation (1 to 2 days). Complete water deprivation may result in death in less than one day in hot weather. Water requirements are greatly increased for body cooling during hot weather. After initiation of the panting reflex, water losses from the lungs are greatly increased.

Chickens are particularly sensitive to poisoning with common salt if it is accidentally introduced in the water or added in excess amounts to feed. Water contamination from cadmium, chromium, copper, mercury and zinc should also be monitored. Using water from wells high in chloride may cause permanent change in hens resulting in poor quality of egg shells.

A condition known as "flushing" can result when hens are deprived of water for short periods in hot weather - overconsumption after access to water is resumed resulting in diarrhea.

Daily Water Consumption Guide for Broilers

No. of Birds Per House	*Age (wks)* / *Temp (°C)*	*1*	*2*	*3*	*4*	*5*	*6*	*7*	*8*
					Liters of Water				
1000	21	30	61	95	132	174	216	254	288
	32	34	98	197	272	356	416	462	473
10,000	21	303	605	946	1325	1741	2157	2535	2876
	32	341	984	1968	2725	3557	4163	4617	4730
20,000	21	605	1211	1892	2649	3482	4314	5071	5752
	32	681	1968	3936	5449	7114	8325	9243	9461
30,000	21	908	1816	2838	3974	5222	6471	7606	8628
	32	1022	2952	5904	8174	10672	12488	13851	14191

Source: North, Mack O., Bell, Donald D., *Commercial Chicken Production Manual.* 4th Ed. 1990

6. Non-nutritive feed additives. Since mixed feeds are universally used in large commercial poultry enterprises, the inclusion of non-nutritive substances as feed additives is a convenient method of insuring a uniform intake. These additives, which are manufactured, distributed and advertised by commercial companies, are sold as concentrates diluted by feedstuff carriers. They include: 1) medications: antibiotics, anticoccidials, insecticides and vermicides (wormers); 2) growth producing factors such as certain organic arsenicals; 3) chemical preservatives such as antioxidants to prevent rancidity in fats; 4) antifungal agents to prevent molds from producing toxic substances; 5) detoxicants such as ferrous sulphate to reduce the toxicity of gossypol; and 6) yellow pigments ("carotenoid products") to color skin and egg yolks when consumers have expressed a preference.

To protect human health all of these additives must pass stringent governmental safety tests before their use is permitted as feed additives.

C. Sources of Poultry Feeds. The aspiring poultry producer faces three possible choices for finding a suitable feed source: 1st) prepare his/her own feeds from locally available grains and supplementing with as many natural feedstuffs as possible; 2nd) mix feeds using locally secured grains plus protein supplements and adding imported concentrates which contain essential mineral and vitamin supplements; or 3rd) rely upon a poultry feed manufacturer to supply nutritionally balanced feed for each class of poultry. If a reasonably-priced mixed feed is available, the small poultry producer may develop a good working relationship with a reputable dealer. Producers may best rely on a mutual exchange and trust developed with a feed dealer. Feed manufacturers may employ a nutritionist to guide in feed formulation. Production comparisons with different diets should be encouraged by the poultry producer and the feed manufacturer. Results should be compared using two or more feedstuffs under identical conditions. More than one trial may be required and statistical comparisons should be considered before firm conclusions are warranted.

D. Feeding Systems. Producers with a few scavenging birds may first want to supplement their birds' diet. Green feeds may be grown in season for birds on range. Chickens, turkeys and other fowl will harvest them directly, or they may be cut, dried and stored for feeding to birds in cages. Green pastures are excellent, high protein, high vitamin, feeds for all types of poultry. The importance of storing water (damming streams, irrigation, winter crops) to keep pastures green and growing out of season cannot be overemphasized. Alfalfa, clover and many local grasses make excellent forage crops. Energy and mineral supplements are usually required to improve performance by range birds. Scavenging or free-range birds can be given grain or concentrate premixes made by mixing grain with some protein and mineral supplements. Concentrates for range fed birds may contain the energy and vitamin and mineral levels of mixed feeds. The protein supplement is left out if the birds are consuming plenty of high protein, growing plants.

Poultry may be fed a small amount of a concentrate feed and then all of a cheap, bulky feed that they will eat. Rice bran with rubber seed meal or palm kernel cake can make an adequate supplement to banana or cassava wastes fed *ad libitum* (free choice).

If the market for chicken meat or eggs is adequate to justify expansion, the producer may consider mixing his/her own feed. There are advantages to using a large number of ingredients in any formula, thus making available a good balance of amino acids and adequate vitamin levels more likely. If commercial vitamin and trace mineral supplements are not available, reasonable levels can often be achieved with local ingredients. Animal by-products, milk by-products like whey and buttermilk, milling products like wheat bran and middlings, all make good vitamin-rich ingredients for poultry feeds. Finally, if available, complete mixed feeds may be purchased.

E. Quality Assurance. More important than the type of any ingredient used is its quality. Corn may be universally recognized as an excellent feedstuff for poultry. However, corn with certain types of mold growing on it can be deadly. The value of any feedstuff should be determined by it's ability to promote growth or egg production. Possible growth inhibiting contaminates in ingredients always need to be considered. These include weed seeds, molds, vermin feces (which may also carry disease with them), toxic chemicals, unusually high levels of one nutrient in a certain ingredient (like high calcium in fish meal), and insect larvae. ***Buyer Beware!***

F. Formulas for Feed Rations. A new poultry producer faced with feeding problems for the first time frequently asks: "What feed formula should I use?" From thousands of successful formulations, there can be no simple answer. Possible feedstuffs vary in content of nutritional constituents, availability in different areas, and cost. Formulas for chickens are often classed as starter, grower, finisher, layer, or breeder rations and they vary with breeds, climate and other classes of poultry. Large manufacturers may change formulation daily or weekly depending upon cost and the availability of substitute ingredients.

Several years of study in poultry nutrition are required to formulate a feed which includes all essential ingredients and is cost competitive. More is known about the nutrition of poultry than any other animal including man. Most large feed companies employ a nutritionist who makes use of computer technology in formulating a *least-cost ration.* Such formulas are often considered confidential. References on feed formulation (Section XI) provide outlines of the essential tools required.

G. Sample Feed Formulas. Beginning poultry producers may want to start with a simplified feed formula like one of those listed below. Ingredients are listed in percentages which may be weighed out in kilos or pounds. If chicks are kept indoors 1/2 kg of stabilized cod liver oil or a commercial vitamin supplement must be added to each 100 kg of feed.

H. Feed Mixing and Delivery. After the feedstuffs required in the formula are assembled, the laborious task of feed mixing can be undertaken. Some ingredients are ground in a hammer mill to provide a fine meal. Bulky ingredients may be poured out on a concrete slab and lesser ingredients, including premixes, are placed on top. Premixes include vitamins, minerals or anticoccidial drugs are used in small quantities. Your own premixes may be prepared by diluting the essential additive with 1-5 kilos of corn or another ingredient. This mixture is blended by turning several times in a closed container (Figs. 59, 60 and 61) or in a plastic bag. All ingredients are turned over several times with a shovel, hoe, or in a feed mixer. As the enterprise grows, mixing the larger quantities of feed may require a mechanical mixer. A hand crank or an electric motor should rotate the drum for several minutes to assure thorough mixing. A small portable mixer works well to mix poultry feed. The mixer should be used only for feed mixing. A separate one is used for mixing cement. Cement dust is toxic to chickens. Rations are usually fed as a dry mash. More sophisticated feed mills use heat and pelleting machinery to produce crumbles or pellets which improve growth and feed efficiency. Although some mixed feed can be stored for a month or two, speed of deterioration of nutrients varies with different ingredients and holding environments. In hot climates some ingredients begin to deteriorate within a week.

Many poultry producers have come to rely upon feed manufacturers to purchase and assemble the ingredients. Specialized heavy equipment is employed to mix and bag the feed. Bulk delivery systems may save expense and labor if storage bins and good roads are available. Unfortunately many producers have been disappointed when substitution has been made in the feed formula by the manufacturer. A loss of reputation is often the best control over the unreliable substitutions sometimes made in a feed mill. Governments sometimes require tags indicating minimal ingredient levels within the bag.

Many aids for transporting the feed to the premises and into the house have been devised. Feed wastage needs to be guarded against. A common rule with a trough-type feeder: “never fill the feeder more than half full”.

Feed Mixers

Fig. 59. Tumble mixer with hand crank. Excellent as a premixer with a 5 gallon can. A 50 gallon oil drum can mix up to 50 kilos of feed.

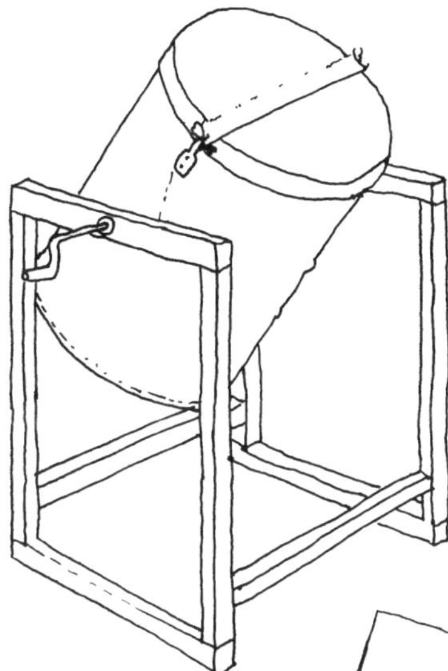

Fig. 60. End view of the same mixer.

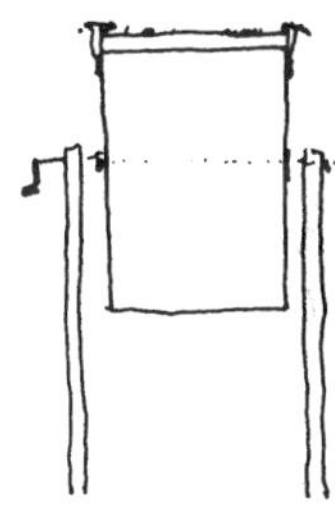

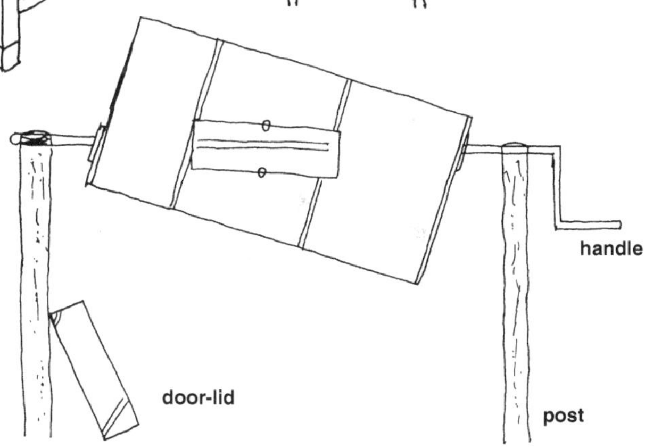

Fig. 61A. Another mounting for a clean 50 gallon oil drum requires a specially fabricated locking door.

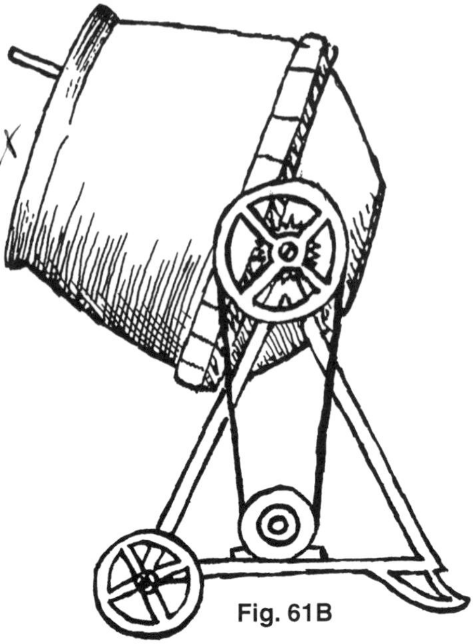

Fig. 61B

Fig. 61C. To prevent waste of feed the lip or top of feeder trough should be as high as the back of the chicken.

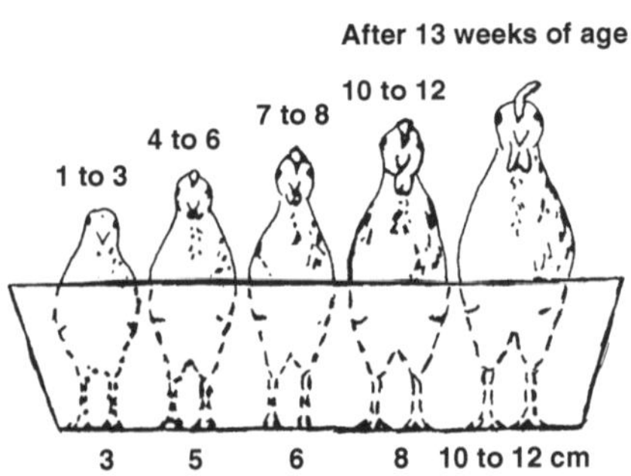

Fig. 61D. Changes in Trough Length with Increasing Age.

Sample Rations for Egg Laying or Dual Purpose Chickens

Ingredient	*Starter A Weeks: 1-8*	*Starter B 1-8*	*Grower 9-20*	*Layers 20+*
	Percentage			
Ground or finely cracked grain	49.5	29.5	49.5	21.5
Grain milling by-products	20.0	20.0	–	25.0
Wheat bran or rice bran	–	15.0	30.0	15.0
Soybean oil or legume seed oil meal	10.0	–	5.0	15.0
Meat, fish, poultry scraps or meals	10.0	10.0	10.0	10.0
Alfalfa or other dried forage	5.0	5.0	–	5.0
Groundnut, sesame or cotton seed meal	–	15.0	–	–
Dried whey or distillers by-product	3.0	3.0	5.0	–
Crushed limestone or oyster shells	2.0	–	–	5.0
Steamed bone meal or feed grade phosphate	–	2.0	–	3.0
Salt	.5	.5	.5	.5
TOTAL	100	100	100	100

Sample Rations for Commercial Broiler Production

Ingredients	*Starter Weeks: 1-3*	*Grower-Finisher 3-10*	*Withdrawal Last week*
	Percentage		
Ground or finely cracked grain	58.65	63.15	68.20
Soybean, or legume seed meal	31.00	22.50	16.50
Corn gluten meal 65%	–	3.00	4.00
Meat, fish, poultry scraps or meals	5.00	5.00	5.00
Fat (feed grade)	3.00	4.00	4.00
Crushed limestone or oyster shells	.65	.65	.65
Feed grade phosphate	1.25	1.25	1.25
Salt	.25	.25	.25
Methionine supplement	.15	.15	.10
Trace mineral mix and vitamin premix	.05	.05	.05
TOTAL	100	100	100

Sample Rations for Laying Chickens fed in Sri Lanka

Ingredients	*Chicks (0-8 weeks)*	*Chicks (8-18 weeks)*	*Layers*
	Percentage		
Tambagalla (sorghum)	39.5	44.5	41.5
Rice bran	7.0	23.0	19.5
Fish meal	10.0	12.0	8.5
Coconut meal	25.0	20.0	18.5
Gingelly (*Sesamun indicum*) cake	12.0	–	2.0
Cowpeas	6.0	–	3.0
Shell grit	–	–	6.5
Salt	0.5	0.5	0.5
TOTAL	100.0	100.0	100.0
Added per 100 kg:			
Potassium iodide (g)	0.145	0.145	0.145
Choline chloride (21.7%) (g)	555	530	540

Sample Rations for Laying Chickens fed in Zaire

Ingredients	*Starter Weeks: 1-8*	*Grower 8-24*	*Layers 24+*	*Layer Concentrate**
	Percentage			
Maize, ground	33.0	40.0	40.0	20.0
Millet, ground	22.0	15.0	20.0	18.0
Rice, dehulled, ground	11.0	–	–	–
Rice, paddy, ground	–	10.0	10.0	10.0
Fish meal	7.5	7.0	3.0	4.0
Meat meal	–	–	3.0	5.0
Skim milk, powder	5.0	6.0	–	–
Yeast, dried	3.0	1.0	1.0	–
Groundnut cake meal	9.0	12.5	12.0	25.0
Alfalfa meal	7.5	5.0	7.0	12.0
Dicalcium phosphate	0.5	1.0	0.5	2.0
Oyster shells	1.0	2.0	3.0	3.0
Salt	0.5	0.5	0.5	1.0
TOTAL	100	100	100	100

*To be fed free choice with a grain.

Sample Rations for Laying Chickens Fed in Uruguay

Ingredients	*Chicks*	*Layers*
	Percentage	
Ground maize	40.5	40.0
Ground wheat	20.0	5.0
Sorghum	–	3.0
Ground barley	16.0	20.0
Bran	–	10.0
Meat meal	15.0	7.0
Ground sunflower cake	7.0	10.0
Oyster shells	1.0	4.0
Salt	0.5	1.0
TOTAL	100	100

Section 7

Poultry Health Programs

A. Disease versus Health (Role of Non-Infectious Causes of Disease):

Useful discussion of poultry disease requires some fundamental understanding of the physiology and biology of poultry. The inexperienced reader is encouraged to learn about the fundamental requirements of poultry in this and other introductory materials (see References). As discussed below, and elsewhere in this book, poultry have numerous requirements for health. If the needs of birds are not met, disease will surely follow. This is true regardless of the actual need or requirement discussed. Nutritional deficiencies lead to nutritional diseases, but may also contribute to the probability of becoming infected with infectious disease agents and increase the likely severity of such infectious diseases. This may often be true even if the nutritional deficiency is not severe enough to cause overt signs of nutrition-related disease. Similar statements can be made for other requirements, including water quality and availability, shelter and protection from extreme weather, physical threats from other aggressive birds or predators, or the presence of toxins in feed. Many common diseases of poultry are caused by environmental problems, toxins, genetic factors, nutritional deficiencies, and trauma, without the involvement, at least in the beginning, of infectious agents. Diseases of toxic, nutritional, traumatic or other non-infectious causes, weaken the immune system and reduce resistance to infectious disease. In some cases, low level disease causing agents which are normally not a problem for poultry can actually infect birds and cause disease because of this reduced resistance. Thus, secondary infections, infections where pathogens are sometimes involved secondarily in diseases of non-infectious origin, may sometimes be observed.

To begin to understand a particular disease problem, the first step is to ascertain the primary cause and nature of the disease. Obviously, merely treating the secondary infections is not the most appropriate means of dealing with non-infectious diseases. Nevertheless, this is sometimes where some experts leave their recommendations for improving the health of poultry flocks, since it is frequently easier to diagnose the secondary infectious disease problem than it is to ascertain the underlying environmental or nutritional causes of reduced immunity and resistance. An improved understanding of the nutritional, toxic, environmental, and traumatic causes of disease, can lead to improved diagnosis and an improved level of performance by addressing the primary causes of many poultry diseases.

When any poultry disease problem is noted, it is critical to examine not only the noticeably affected birds, but also other birds within the flock, preferably in the environment where they are maintained. The environment must be carefully examined for appropriate housing and shelter, space and general cleanliness, availability of good quality water, and proper and available toxin-free nutrition. As some breeds are more suited to certain environments, breed selection (genetics) should also be considered. All too frequently, specific infectious diseases are blamed for diseases that in actuality could be avoided by proper management and nutrition, as covered elsewhere in this book. Nevertheless, infectious diseases, whether primary or secondary causes of disease, play an important limiting role in many poultry production systems.

B. Infectious Dose (Dilution is the Solution to Pollution):

For most diseases caused by infectious disease-causing agents (pathogens), the bird must be exposed to a relatively large number of organisms (pathogens) for infection to occur. This number of pathogens is often referred to as the "minimal infectious dose". Obviously, the minimal infectious dose changes with age of the bird, general health and condition of the bird, the amount of environmental stress, nutritional quality, and many other factors that are sometimes difficult to measure. The main point to be gained is that a large number of most disease causing organisms (pathogens) are required to cause infection. Furthermore, for a given pathogen, the severity of disease in an individual animal may be related to the number of pathogens to which the animal was initially exposed. Thus, sanitation, the process of reducing the number of pathogens, has a critically important role for poultry production in spite of the fact that sterility (the total absence of organisms) can almost never be achieved under real world conditions. This gives credence to the statement, that with regard to infectious diseases, "dilution is the solution to pollution". If we can dilute the pathogen below the infectious dose, infection simply will not occur. Other times, perhaps a reduction in the number (concentration) of pathogens can be achieved but it is still above the infectious dose. In this case, even though the birds become infected, the severity of the disease may be reduced by exposure to a lower infectious dose. Maintaining a clean environment, rotating areas where birds are held (movable cages), and frequently replacing or changing bedding material, are frequently simple techniques that will reduce the chances of significant infections.

In some cases, protecting birds from an infectious level of a specific disease causing agent is not possible. As an example, Marek's disease (discussed below) virus is shed in the feather follicle dander of infected birds and is readily transmitted to susceptible chickens at doses which are infectious. In this instance, cleanliness of the environment is not likely to prevent infection and disease. In these cases, it is necessary to increase the resistance of the bird by increasing specific immunity through the use of an appropriate vaccine. In effect, vaccines (discussed below) increase the number of a specific disease-causing agent which are required to infect and cause disease. It is therefore important to remember that there are three important considerations for prevention of infectious disease in poultry. First, proper nutrition and management to minimize stress is essential. Second, every effort should be made to reduce exposure to disease causing agents (as discussed above). Third, specific appropriate vaccines can increase the resistance to challenge by a specific disease-causing agent. In healthy poultry, infectious disease can be prevented by maintaining resistance above the level of challenge, or by reducing challenge below the level of resistance (immunity).

C. Sanitation and Biosecurity: Isolation of Poultry from Infectious Disease Causing Agents

There are two distinct sanitary principles that allow for maintaining large numbers of poultry for food production. While there are many individual practices that can contribute to overall sanitation and biosecurity, these two principles should be observed whenever possible.

1) Strict Isolation: If new pathogens are not introduced, the number of infectious disease problems will be limited to those pathogens present in the environment or carried by the birds. The original concept of **all-in, all-out** poultry production is the basis for modern practices which allows intensive poultry agriculture. By limiting the flock origin to a single source obtained at a single time (all the same age), pathogen exposure is greatly reduced. Because many specific pathogens require a host for long-term propagation and even survival, eliminating multiple species, ages, and flock sources of birds simultaneously present on a premise has greatly reduced the introduction and persistence of pathogens. A common breach of isolation and all-in, all-out practices occurs when growers allow a few birds to remain in the production facilities during "down times" between flocks. This allows for maintenance of a reservoir of pathogens in the remaining birds. It is absolutely essential to make use of any opportunity to depopulate the flock and begin anew with healthy chicks from a single source. In this way, old problems need not necessarily remain problems indefinitely.

While there are a number of biological vectors or carriers of diseases including poultry, wild birds, rodents, some insects, and even other farm animals (not common), the most important controllable source of disease is other poultry and people that travel between flocks. Additionally, animals (including humans), equipment, and machinery can serve to mechanically transmit disease causing agents. For example, people visiting multiple flocks may unintentionally transmit disease causing organisms on clothing, boots, and tools carried onto a premise. Similarly, neighbors, dogs, cats, wild animals, and other sources may serve to unintentionally transmit disease causing organisms. Since every contact contains risk, reducing the number of contacts will reduce risk. Furthermore, appropriate sanitation practices (please see below) will also serve to reduce the risk of fomite introduction.

Fig. 62. Large roundworms and cecal worms.

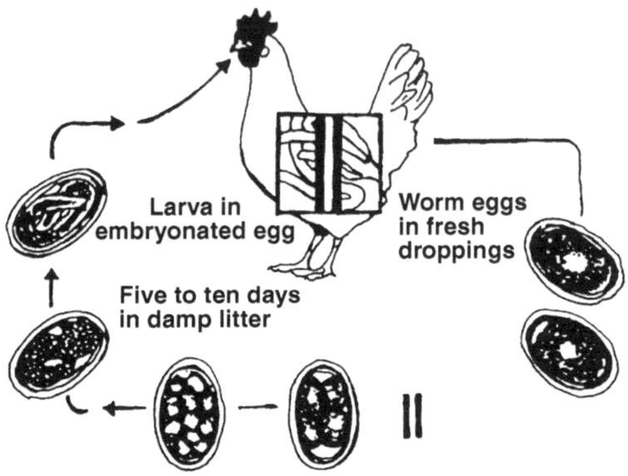

Break the life cycle ‖ by treating the birds to remove adult female worms or remove infective embryonated eggs in litter

Fig. 63. Life Cycle of Coccidia and Severity of Infection.

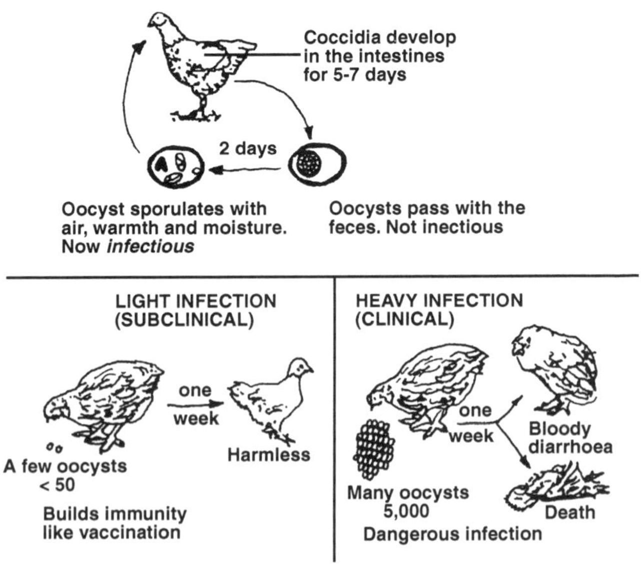

2) Reduction of Pathogen Numbers: In most cases, exposure to a relatively large number of infectious organisms is required to infect a susceptible host as described above. This number of infectious pathogens of a single specific type is known as the **minimal infectious dose.** This number will vary widely depending on a variety of factors such as the general health and age of the host and the specific virulence of the pathogen. For poultry production, we can exploit this knowledge to use sanitation procedures whenever possible. Even though we know that the environment can rarely be sterilized (all organisms killed), we can often take steps to reduce the number of infectious or potentially-infectious agents in the environment. This process is known as **sanitation.** Pathogen numbers in the environment can be reduced by physical removal (such as clean-out, replacement of used litter, and physically washing) or by killing pathogens in place (disinfection). Usually these procedures work best when used together, physically cleaning the environment prior to the application of chemical disinfectants. When one realizes that physical cleaning and disinfection can sometimes reduce the number of pathogens below the minimal infectious dose, one realizes that "dilution can truly be the solution to pollution" when it comes to infectious disease problems. There are many debates as to the best disinfectant for a specific purpose. However, there is no debate as to the benefits of physically cleaning away infectious material. As a final step, any disinfectant (including a dilute household bleach solution) is usually quite effective.

In addition to physically cleaning and disinfecting the environment, time also tends to work toward reducing pathogen numbers in the environment. Most of the important pathogens of poultry are **obligate pathogens,** meaning that these organisms cannot replicate in the absence of the poultry species which they infect. Thus, as we remove all poultry from the facility, these organisms tend to decline over time. The survival of different pathogens, of course, is highly variable, with some (e.g. Fowl Pox) persisting for many years while others (e.g. *Mycoplasma spp.*) are very fragile in the environment and disappear in a matter of hours after poultry are removed. Environmental conditions can also influence pathogen half life, with higher temperatures, high ammonia levels, drying, sunlight, and chemical disinfectants tending to greatly reduce the survival time, thereby accelerating death of pathogens. This phenomenon is exploited in commercial poultry operations using the all-in, all-out production concept described above. The time between flocks is important for reducing pathogen levels within the environment. In commercial poultry operations, veterinarians and field service-persons will sometimes elect to extend the "down time" between flocks to more fully exploit this phenomenon, particularly when faced with disease causing agents that decline rapidly over time. In general, it is recommended that the premise contains no poultry for at least 2 weeks (after cleaning and disinfection) before arrival of new birds.

D. Protection by Vaccination:

Vaccination is the process by which specific immunity can be induced by exposing the host to a weakened or killed organism that has the same external "appearance" as the field pathogen to the immune system. All specific disease-causing agents have structures on the organism's surface that appear quite unique to the bird's immune system. These structures which can be recognized by the immune system (epitopes) are generally unique for a specific disease-causing agent. By exposing the bird's immune system to this vaccine **prior** to infection, the chances of infection and/or disease can be greatly diminished. The key to successful vaccination is to administer the vaccine before the animal is infected. After infection, vaccinations for that disease will not help the birds recover and, in some cases, may hurt the birds. Vaccine selection is based on the history of infectious disease problems in a given premise and the prevalence of a certain disease in a given area. It should be noted that there are some diseases which have worldwide prevalence and which vaccination programs are always recommended (e.g. Marek's disease).

1) Importance of Vaccine (Serotype) Selection: In some cases, a single disease agent can have alternative "appearances" to the bird's immune system (multiple serotypes). In this case, it is essential that the vaccine chosen appears the same to the immune system as the actual disease-causing agent to which the birds are likely to be exposed in order to induce protective immunity. Because of this problem, not all vaccines for a specific disease problem are capable of protecting against all organisms of the same disease name. In these cases, the specific disease-causing agent must be isolated in a diagnostic laboratory, and the serotype determined, in order to choose an effective vaccine for present or future flocks. The fact that some disease causing agents can present as one of a large variety of serotypes (appearance to the immune system) is the reason that vaccination is not usually useful for those diseases (e.g. there are a large number of possible serotypes for Avian Influenza and vaccination for this disease is only done under very unusual circumstances). However, for small flocks in primitive areas it is recommended that chicks are vaccinated against Marek's Disease, Newcastle Disease, and Infectious Bronchitis at the source hatchery if possible. Only under specific circumstances, beyond the scope of this book, are other specific vaccines likely to be useful.

2) Relationship of Vaccination to Infectious Dose: It is very important to remember that protection by vaccination is also related to the concept of infectious dose, discussed above. With exposure to higher infectious doses, immunity can often be overcome. Thus, the degree of immunity or protection is relative. Vaccination usually provides protection against a higher challenge level (exposure to an increased number of a specific disease-causing agent). If the induced immunity is sufficient to protect against the challenge to which birds are likely to be exposed to in the field, then the birds are most likely protected against that disease. The simple fact that a flock was vaccinated, therefore, does not necessarily mean that the flock is absolutely protected from a given disease. Additionally, there may be other serotypes of a given disease against which a specific vaccine will offer no protection. Thus, it is very important to maintain strict isolation and sanitation even in vaccinated flocks.

E. Field Identification of Infectious Disease Problems

Treatment of infectious disease problems with antibiotics is frequently a "last ditch" effort in the absence of a specific diagnosis and knowledge of susceptibility of a particular agent to the chosen drug. Avoiding infectious disease problems of poultry is infinitely better than treatment in all circumstances, and particularly under primitive conditions where diagnostic support is not available. Diseases caused by viral or fungal agents are not responsive to commonly available antibiotics which can be used in poultry. Similarly, roundworms, tapeworms, lice, mites, and protozoal pathogens (e.g. infections with coccidia) all require specific treatments and are not susceptible to common antibiotics. Diseases caused by specific bacterial pathogens can sometimes be successfully treated if an appropriate antimicrobial drug is selected. However, many isolates are resistant to common antibiotics and these may not respond to selected drugs unless (as determined by a diagnostic laboratory) that specific isolate is susceptible to a particular drug. As competent diagnostic services are frequently unavailable to producers of small flocks in remote locations, it is very difficult to correctly guess which drug might be effective. Far too much emphasis is generally given to the use of antimicrobial drugs in the maintenance of healthy poultry, particularly those raised in small flocks far from diagnostic services. Nevertheless, it is sometimes necessary to identify and categorize disease problems for the purpose of appropriate preventative measures for future flocks, and for the treatment of bacterial infections when there is no other hope.

1) Bacterial Pathogens: Bacteria are ubiquitous (found almost everywhere) in the world and there are literally thousands of bacterial species that have been characterized. Most bacteria in the world are

Anatomy of Chickens

Fig. 64. External features.

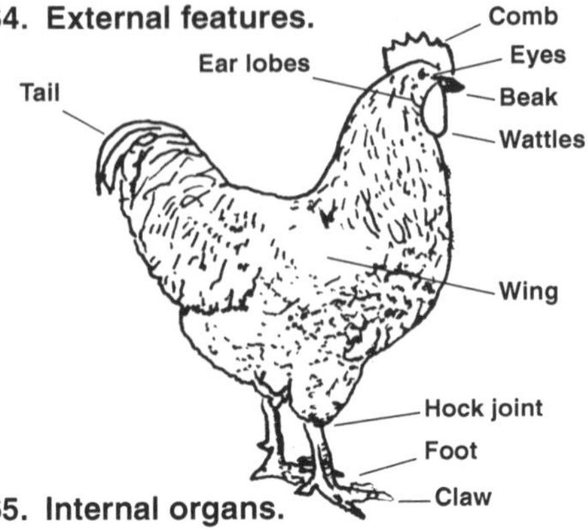

Fig. 65. Internal organs.

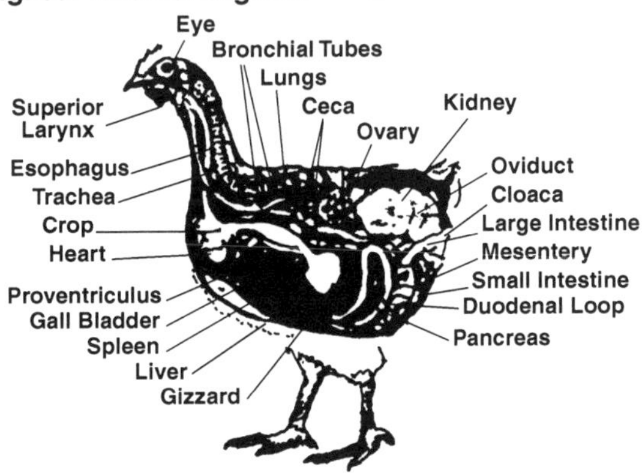

Fig. 66. Digestive system.

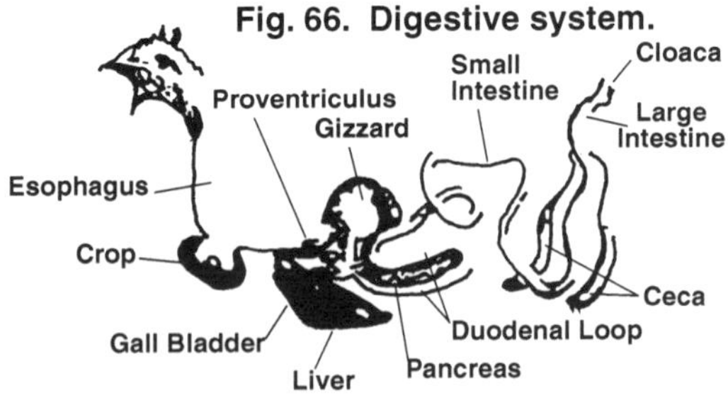

not capable of causing disease in poultry, others only cause disease when an opportunity (e.g. immune suppression, concurrent disease, wounds) allows them to infect the bird. A third group of bacteria are primary pathogens and propagate mostly or solely in the host bird. These bacteria are known as **obligate bacterial pathogens** and generally constitute the most virulent pathogens of poultry.

Bacterial pathogens often produce toxins which cause leakage of blood from tiny blood vessels (hemorrhage) which can often be observed against the light background of the abdominal fat, fat surrounding the heart, and sometimes in the visceral organs. Also these toxins attract the cells that form pus to the affected sites. Thus, the presence of multiple hemorrhages and accumulations of pus is often an indication of involvement of a bacterial pathogen, which could be treated with an appropriate antibiotic.

2) Viral Pathogens: Viral pathogens differ from bacterial pathogens in a number of ways. For practical purposes, diseases caused by viruses are less likely to cause the multiple hemorrhages observed in bacterial infections and are less likely to cause the accumulation of the large numbers of cells that form pus. Thus, infectious diseases which do not cause widespread hemorrhage and accumulations of pus are most likely to be of viral causes.

Prevention of viral infections of poultry is the only real mechanism for avoiding the related disease and production losses. Prevention is based on one of 2 mechanisms: avoiding exposure of the flock to a specific pathogen (sanitation and biosecurity) or through immunization. The reader is reminded that general nutrition and management contribute to susceptibility and severity of infections by many viral disease agents.

3) Fungal Pathogens: Yeasts, molds, and mushrooms are in the family that are called fungi. These are very simple cells which share a number of characteristics with cells of poultry. Thus, it has proven difficult to develop cost-effective antifungal chemotherapeutic agents for use in commercial poultry.

Because fungal pathogens are usually not affected by common antibiotics used in commercial poultry, these antibiotics may actually allow the opportunity for pathogenic fungal infections by eliminating bacterial flora (normal non-pathogenic bacteria) that usually compete with the fungi for nutrients and space (e.g. Crop Mycosis). Indeed, most fungal infections of poultry are opportunistic. For example, the severe respiratory disease called pneumomycosis (brooder pneumonia) is caused by the inhalation of tremendously high numbers of fungal spores (often *Aspergillus fumagatus*) due to moldy conditions in the environment.

As there are no really effective treatments for fungal infections of poultry, the only cost-effective strategy for avoiding these disease problems is avoidance of the conditions that predispose to fungal infection.

It is important to note that mycotoxins (*myco* = fungal) are preformed toxins generally produced by fungal species that are non-pathogenic for poultry. These species may infect grain-bearing plants such as corn and produce toxins even prior to harvest. Mycotoxins are generally highest in years of high stress (e.g. drought) for the plants, supporting the concept that many of these toxins are produced even before harvest. However, the greatest problem with mycotoxin generation probably occurs after harvest during grain storage and shipment. Warm moist conditions are particularly suitable for mycotoxin-producing fungi to grow and contaminate grain sources. Generally grain moisture contents less than 14% do not support mycotoxin-producing fungal growth.

In some warm areas with high humidity, particularly with sub-optimal grain storage facilities, mold inhibitors are added to grains and finished poultry feeds to retard the growth of fungi and production of mycotoxins (see chapter dealing with grain storage). It is important to note that commercially available mold inhibitors do nothing to remove existing mycotoxins, they only prevent continued growth and production of toxins by the fungi.

F. Chemotherapy

Treatment of infectious disease problems of poultry is often disappointing due to resistance by some bacterial pathogens against selected antibiotics. Nevertheless, treatment is sometimes useful and indicated. Specific bacterial disease problems which are usually responsive to antibiotics include Mycoplasmosis (tetracyclines, tylosin) and Necrotic Enteritis (bacitracin, streptomycin, neomycin, or penicillin). Common bacterial diseases such as Fowl Cholera, Colibacillois, and Salmonellosis, are frequently resistant to common antibiotics, and laboratory testing of the isolate to determine sensitivity is strongly recommended. It is important to remember that antibiotic treatment is a game of selective toxicity, trying to poison one organism without harming the other. Therefore, inappropriate treatment may actually cause harm rather than simply not causing improvement. Also, antibiotic treatment can reduce the beneficial normal bacterial flora, increasing susceptibility to resistant bacteria and fungi, as mentioned above. To minimize detrimental effects of antibiotic usage, it is important to follow manufacturer's recommendations precisely.

G. General Vaccination Considerations

Vaccines are sometimes expensive, laborious to administer, and sometimes themselves cause low level disease problems referred to as "vaccine reactions". Thus, it is important to select the minimal vaccination strategy that will protect commercial poultry flocks from the specific disease challenges they are likely to encounter. Because a second, or even a third, immunization is sometimes required to generate protective immunity, it is important to determine the appropriate vaccines for the environment where birds are raised, based on geographical area and the likely pathogen challenges poultry are likely to face. In some areas, successful vaccination strategies will necessarily be much more intense (expensive) than in other areas. The intended use of the flock is also an important consideration, with flocks intended for breeding or table egg production receiving considerably more frequent boosting (re-vaccination) than those intended for slaughter at 6-7 weeks-of-age.

A common vaccination strategy for broiler chicks includes Marek's disease vaccine at 18 days of embryonation (injection of the egg) or at day-of-hatch (injection) and spray administration of Newcastle Disease and Infectious Bronchitis in the hatchery (spray box). Infectious bursal disease vaccines are available for application at day-of-hatch. However, many field isolates will "break" higher levels of maternal antibodies than the killed vaccines. Generally, it is much preferable to obtain chick from breeder flocks with high and consistent titers against Infectious Bursal Disease Virus (Gumboro). Attenuated live vaccines are commonly available for use at 10-16 days of age (drinking water) as maternal immunity begins to wane. These vaccines are recommended when Gumboro problems are occurring around 21 days of age.

In some areas young pullets are also vaccinated for Laryngotracheitis, Avian Encephalomyelitis, Infectious Coryza, Fowl Cholera, and Fowl Pox. Re-vaccination (boosting) for Newcastle Disease and Infectious Bronchitis is also common in chickens intended for egg production. Specific recommendations vary widely with area and selected vaccines. In small isolated flocks which practice good biosecurity and all-in, all-out production, vaccination programs can be greatly simplified and little or no vaccination beyond the hatchery may be required. As specific disease problems are recognized in an area, it becomes prudent to incorporate an appropriate vaccination strategy for future flocks. Always follow manufacturers directions for using vaccines and consultation with animal health workers is sometimes useful to learn what vaccines are being used and appear to be working in a given area.

H. Disease Diagnosis Considerations

1) Importance of Accurate Diagnosis: While the information discussed above provides useful generalized information, solutions to specific problems require an accurate diagnosis. In some cases, accurate diagnoses may be difficult even for experienced poultry veterinarians without diagnostic laboratory assistance. Since this is not always possible, it is sometimes useful to analyze the available data and to formulate a "presumptive" or "field" diagnosis. Diagnoses of this type are best made using a process of elimination, rather than trying to "fit" a particular disease agent to a specific case. Most veterinarians in the field try to determine the type of etiology (bacteria, virus, fungus, toxin, or parasite, etc.) and identify the organ systems affected. In this way the possibilities can be quickly narrowed without overlooking important clues. The following discussion may assist in field diagnosis. Whenever possible, confirmation with a diagnostic laboratory can help with future diagnoses and future poultry flocks by anticipating problems.

2) Recognition of Infectious Disease Type: After determining that a disease problem is infectious in nature, the next step in the diagnostic process is to develop a preliminary diagnosis of the type of infectious agent involved. This is sometimes very difficult without laboratory support. Nevertheless, many times a useful presumptive diagnosis can be made in the field, hopefully later to be proven correct based on laboratory testing. The first step is to determine if the infectious agent involved is a virus, bacterium, or fungus (internal and external parasites are discussed below).

When we remember that many viruses cause disease by destroying cells during viral replication, it is easy to realize how viral diseases may look similar. Bacterial pathogens, on the other hand, have a more complex array of means to damage the host, often resulting in a larger variety of lesions produced. Bacteria can produce a variety of toxins which frequently leave telltale signs of their involvement. For common poultry diseases, these bacterial toxins are potent attractors of heterophils, cells serving as the first line of defense (phagocytes). Large numbers of heterophils are observed grossly as "pus", and the presence of purulent (pus-containing) exudates are often a strong clue that a bacterial pathogen is at work. Also, some of these toxins cause capillary fragility and rupture, leading to hemorrhage. These tiny hemorrhages, referred to as petechial hemorrhages, are often best observed against the relatively light background of fat tissue at necropsy. Most commonly, these hemorrhages will be observed in the abdominal fat pad or the pericardial fat.

Fortunately, there are only a few fungal pathogens of poultry, and the most common of these diseases (Crop Mycosis, Aspergillosis) usually present with clinical signs and lesions that are peculiar to these diseases.

Viral-caused diseases often cause inflammation with an absence of large amounts of pus or systemic petechial hemorrhages. As virus replication causes cell death and tissue destruction, tissue erosions and ulcers are sometimes observed. Congestion of blood vessels without hemorrhage is often a lesion associated with viral pathogens. When combined with an absence of pus, viral-etiology is usually the best bet.

3) Recognition of Affected Physiological Systems: Another way of reducing the number of potential diseases which could be causing the problem is to determine the system(s) affected. As the classification (toxic, nutritional, infectious, etc.) of the disease problem has been established, the involved system(s) should be considered. Many diseases are relatively specific with regard to the system(s) primarily affected. For example, some nutritional diseases primarily affect bone strength (e.g. Rickets, Osteomalacia) and some affect skeletal muscle and brain function (e.g. Hypovitaminosis E, selenium deficiency). Alternatively, some viral diseases primarily affect the respiratory tract (e.g. Infectious Bronchitis), others affect both the respiratory and gastrointestinal tract (e.g. avian influenza and Newcastle disease), while others cause most observable disease at the level of the intestinal tract alone (e.g. Hemorrhagic Enteritis of turkeys). Mental organization of diseases according to the system(s) affected is an excellent way to reduce the number of possible diseases that must be considered (called differential diagnoses). Therefore, the following text on specific diseases are not only grouped by pathogen type, but also by system(s) affected.

4) Recognition of Body Condition: Often the body condition of poultry is indicative of the duration of a particular disease problem and, obviously, this characteristic is of paramount importance when considering affects on production efficiency. Meat-type chickens and turkeys have "normal" breast muscles forming a pronounced convex curvature when evaluated from the keel to the ribs laterally. Leghorns (egg-type chickens) normally have a breast forming more of a straight line from the ventral aspect of the keel to the ribs. In broilers and turkeys, loss of this normal convexity of the breast in broilers and turkeys is suggestive that the animal has begun using muscle tissue for energy purposes and is thus starving (remember, starvation can occur even with the presence of feed). Similarly, the development of a concavity of the breast muscle (from the ventral aspect of the keel laterally to the ribs) is strongly suggestive of early starvation in egg-type chickens.

The amount of body fat is also a very useful indicator of the severity and duration of starvation. Usually, with recognizable changes in breast muscles there is a pronounced reduction of subcutaneous and abdominal fat. As starvation continues, the last area of fat to disappear is that associated with the heart (pericardial fat) in poultry. When an absence of fat tissue is noticed on the heart, the starvation condition is severe.

5) Disease Field Diagnosis: It should be remembered that individual diseases may have variable presentations depending on the specific strain of the pathogen involved, the particular stage of the disease when the birds are examined, or the presence of multiple (complicating) infections. It is therefore sometimes helpful to re-examine the flock several times as more characteristic lesions may be present.

As some diseases may affect different systems at different stages of the disease, or depending on the actual isolate involved, some diseases are listed under multiple organ systems (below).

I. Diseases Affecting the Nervous System

Marek's Disease: A viral disease of chickens which causes nerve damage and tumors (skin and sometimes internal organs). Disease is often observed with a peak incidence early in life (1-3 months of age) and is sometimes called "7-11" disease because peak incidence is more or less observed at these weeks-of-age. Affected birds often present with paralysis of a wing or leg, with one leg dragging or a wing drooped. At necropsy, peripheral nerves will be enlarged (different sizes on each side). Tumors associated with feather follicles, or less frequently internal organs, may or may not be present. Prevention in future flocks through vaccination in the hatchery is required.

Botulism: A toxic disease of all birds caused by ingestion of pre-formed toxin from rotting or decaying flesh, insects, or vegetation. Birds present with flaccid (limp) paralysis, often referred to as "limber neck". Feathers may be unusually loose, but no other lesions are observed. Prevention by identification and removal of the source of toxin. Prompt removal of dead birds is important as the causative bacteria can grow in dead carcasses and the toxin can be concentrated in maggots, which are attractive to poultry.

Epidemic Tremor: Viral disease also known as Avian Encephalomyelitis which affects chicks and poults. Affected birds show tremors of head, neck ,and legs. Holding birds upside down accentuates the tremors. Chicks from immune flocks may be resistant. Specific vaccines are available.

Vitamin E/Selenium Deficiency: Disease is most often observed in chicks and poults less than 2 months of age. Ataxia, falling backward, and loss of balance is responsible for the name "Crazy Chick". Hemorrhage and softness of the cerebellum can sometimes be visually noticed if the brain is dissected from affected chicks. Correction of the deficiency is prevention and treatment.

Aspergillosis: The fungus causing respiratory aspergillosis can sometimes infect the brain, causing nervous system signs. Respiratory disease is more easily diagnosed (below).

Fowl Cholera: The bacterium causing this respiratory disease sometimes affects the cranial bones and the brain, causing nervous signs. The respiratory disease and chronic forms are more easily diagnosed (below).

J. Diseases Causing Tumors

Marek's Disease: This viral disease causes easily diagnosed signs of paralysis (above) and also tumors of the skin, and most commonly the visceral organs. High incidence in young birds (1-3 months) helps to differentiate from Lymphoid Leukosis. Prevention is by vaccination in the hatchery (see section IV, below).

Avian Leukosis Viruses: A slowly developing viral cancer of chickens (related viruses may affect turkeys). The virus may be passed vertically from infected parent flocks or introduced into flocks. Visceral tumors are noted in emaciated birds, often associated with the liver or kidneys. Disease tends to be low incidence within the flock for many months with occasional deaths. This slow and chronic disease, with the absence of nervous signs, helps to differentiate from Marek's Disease (above). No vaccines are available.

K. Diseases Affecting the Musculoskeletal System

Nutritional: Malnutrition, through absence of specific nutrients, limitation of feed access, poor quality feed, or diseases which interfere with food consumption, digestion, and/or assimilation, cause the loss of body condition associated with wasting (as discussed above). In addition to the nervous system signs discussed above, specific deficiencies of vitamin E/Selenium can cause muscle weakness, characterized by white streaks or masses in the breast and thigh muscles at necropsy. Similar lesions can be caused by deficiencies of sulfur containing amino acids.

Malabsorption: Several viral diseases have been associated with malabsorption of nutrients in poultry. More commonly, malabsorption may be caused by moderate to severe coccidial infections as described below. Appearance of undigested feed (feed passage) in feces is a sign of malabsorption.

Mycoplasma Infection: Some forms of this bacterial disease cause swelling of joints and tendon sheaths, often most apparent in the hocks. Fluid from joints and sheaths is sticky and has been described as a honey-milk color. Treatment with antibiotics such as tylosin or tetracycline will improve the condition (this bacterium does not develop resistance easily). Maintaining Mycoplasma-free flocks is much preferable to treatment. As this disease can be transmitted from parents through the egg, it is important to obtain from mycoplasma-clean breeder flocks in all-in, all-out production.

Viral Arthritis: Disease causing swelling of the tendons and tendon sheath, most apparent just above the hock. Hock joints, tendons, and tendon sheaths may contain bloody exudate and the hemorrhage may be found in the tendon itself. Tendons sometimes rupture, causing the bird to be unable to extend the hock joint. Vaccines are available. Chicks should come from reovirus free breeder flock origin for use in all-in, all-out production.

Bacterial Arthritis: In addition to some *Mycoplasma* infections (above), essentially any bacterial infection that moves throughout the body through the blood (septicemia) can cause joint infections and chronic lameness. When this occurs, more than one joint will commonly be affected, although the hock joint is often the most easily observed site of problems. Often the lameness and joint swelling associated with septicemic bacterial infections becomes apparent after the initial cause of infection begins to resolve. Common examples include the bacterial causes of Fowl Cholera, Bumblefoot, Salmonellosis, Respiratory or Enteric Collibacillosis, or even bacterial infections secondary to severe Coccidiosis. Bacterial infections of the joints are difficult to treat as many antibiotics to not enter the areas of infection and treatment is frequently not satisfactory in poultry. On occasion, antibiotic therapy may be attempted if the causative agent and antibiotic sensitivity profile is known.

Chronic Fowl Cholera: This disease most often occurs in the latter stages of the respiratory or "acute" forms of Cholera (see section V, below), as birds begin to recover. Commonly characterized by swellings of joints, wattle, foot pads, tendon sheaths, etc. Exudate at necropsy contains pus and appears cheesy, and is sometimes seen in the eyes (conjunctival sacs) or in the sinus just below the eyes (infraorbital sinus). Nervous system signs (twisting of the neck, holding head upside down or "star gazing") is sometimes observed as the bones of the head are sometimes involved. Antibiotics may sometimes lessen the severity of these lesions but treatment is generally unsatisfactory as these lesions and problems may persist for long periods.

L. Diseases Causing Immunosuppression and General Conditions

Marek's Disease: This viral disease of chickens is spread by feather follicle dander and chicks are usually infected early in life. Vaccination at the hatchery or as early as possible is recommended for all chickens. Virus causes paralysis as transformed lymphocytes infiltrate peripheral nerves, and individual birds are sometimes observed with a drooped wing or dragging a leg. At necropsy, enlarged nerves can sometimes be identified and can be noticed by comparing for bilateral asymmetry. These cells also occasionally infiltrate the iris causing a light color and sometimes blindness. Feather follicles are often enlarged and reddened. Individual tumors within the viscera are perhaps most common but affected birds may present with any of the processes mentioned above. This suppresses the immune system of affected birds and contributes to a host of secondary disease problems. Prevention is by vaccination. Particularly virulent problems not controlled by common vaccination (turkey herpesvirus, serotype 3) may require selection of an alternative vaccine.

Infectious Bursal Disease (Gumboro): Immunosuppressive disease of chickens (mostly 3-6 weeks of age) which causes diarrhea and dehydration and picking/self mutilation of the vent and tail head area. Immune suppression may cause life-long increased susceptibility to other infectious disease problems. Field diagnosis is made at necropsy by examination of the bursa of Fabricius which is inflamed and sometimes hemorrhagic during the early infection, promptly followed by atrophy and reduced size of the bursa. Vaccination may not be necessary in areas where this disease problem has not been identified. Vaccination of breeders can protect chicks for several weeks. Several vaccines are available for vaccination of chicks in areas where this disease has consistently been identified, and may be administered by injection at hatch or in the drinking water or eye drop between 1 and 2 weeks-of-age. Selection of effective vaccination programs has been difficult and may require correspondence with a diagnostic laboratory or poultry veterinarian.

Mycotoxicosis: Mycotoxins are poisonous compounds produced by certain fungi. There are more than 200 known mycotoxins which can affect poultry. The most common mycotoxin is Aflatoxin which causes aflatoxicosis, characterized by hemorrhages throughout the body, pale fragile livers (due to lipid infiltration), and immune suppression. Other mycotoxins cause renal stones and dysfunction (ochratoxin), hemorrhagic burn-like mouth lesions (T2 toxin), and many other less common conditions. While there are some feed additives that can reduce the effect of some specific toxins, these additives are probably of little value in non-commercial poultry production. Prevention of fungal (mold) growth by proper drying and storage of grains is critical for preventing the production of these important toxins after harvest (please see chapter in this book regarding grain storage). It should be noted that some important toxins, such as aflatoxin, can be formed in high concentrations in the field prior to harvest of some grains. Conditions that stress plants (e.g. drought) in the field increase susceptibility of the plants to parasitism by the fungi that can produce mycotoxins, and toxins can be present even before storage. In areas where storage of grain or complete poultry feed without moisture contamination is difficult, commercial products to reduce mold growth during storage are available. However, it should be noted that these mold/fungus-inhibiting products do not detoxify feed which has already become contaminated.

Fig. 67. Bamboo cage house with a thatched roof. Note the protective barriers on the legs to keep out rodents and snakes.

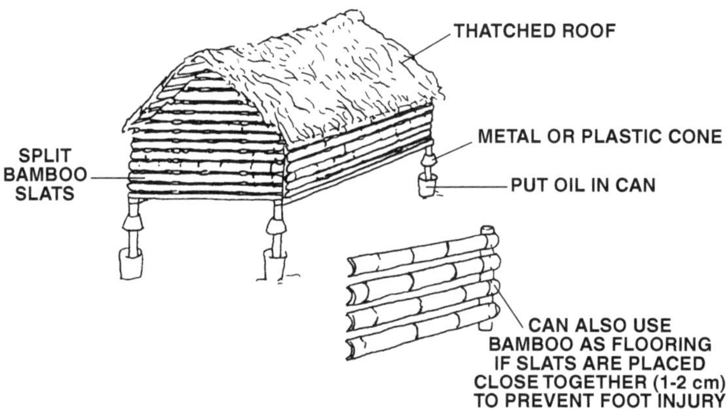

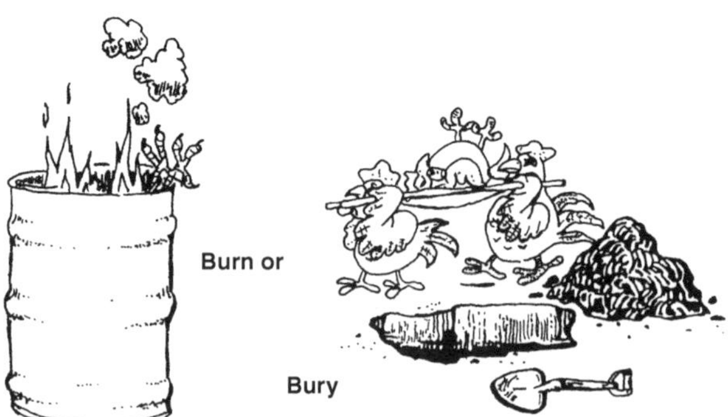

Fig. 68

M. Diseases Affecting the Respiratory Tract

Conditions Involving the Eyes and Face

Fowl Pox: This viral disease can be transmitted many miles by certain mosquitoes and perhaps other blood sucking flies to previously uninfected flocks. The virus is very stable in the environment and difficult to eradicate from a premise once infected. Within a flock, the virus tends to spread from bird to bird slowly by direct contact with injured or abraded skin/mouth areas. Affected birds may have lesions on the unfeathered areas of the skin resembling tumors which may or may not be associated with scabs, depending on the stage of infection. Tan colored plaques may also be formed in the mouth or near the larynx of birds infected by picking at skin lesions of other birds. Area around eyes are often involved. Affected birds will completely recover unless debilitated by other secondary infections or starvation occurs due to inability to find feed/water or due to pain associated with eating/drinking. Prevention is by vaccination, usually applied by a two-pronged applicator which is pushed through the wing web of chickens (thigh of turkeys). Effective vaccination produces a scab or "take" at the vaccination site about a week after administration of this live vaccine. Vaccination of subsequent flocks may be necessary as this virus tends to remain in the environment.

Infectious Coryza: Caused by a bacterial infection (*Hemophilus paragallinarum*) in chickens, pheasants and guineas. Infection primarily affects upper respiratory tract and face, causing generalized facial swelling (edema), filling of the sinus below the eyes (infraorbital sinus) with hard, cheese-like pus, and the expected sneezing and nasal discharge. Infected birds are more susceptible to secondary infections such as *E. Coli* or *Mycoplasma* which may also affect the lower respiratory tract, complicating diagnosis. Treatment with antibiotics to which the organism is susceptible may reduce the severity of disease, but will not eliminate the infection (e.g. Tetracycline, spectinomycin, etc.). The causative organism is very fragile in the environment outside of the host and is eliminated in a few days between flocks in all-in, all-out production. Replacement with Coryza-free birds will eliminate the disease in future flocks unless reintroduction is allowed due to poor biosecurity.

Fowl Cholera: The chronic form of Fowl Cholera (discussed in section III, above) can cause facial swelling, ocular and nasal discharge, and pus formation in the wattles as discussed. The more acute respiratory form, which precedes chronic Fowl Cholera, is discussed below.

- Limited to the Upper Respiratory Tract

Infectious Coryza: Facial swelling, nasal discharge, filling and distention of the sinuses below the eyes (infraorbital sinuses). Details are discussed above.

Infectious Bronchitis: One of the most highly contagious respiratory viruses of chickens. Primarily affects the upper respiratory tract, but causes marked declines in egg production and misshapen eggs with thin albumen. In chicks, may cause mucus plugs which cause asphyxiation, especially if chicks become excited. Mucus plug may be observed in the trachea of young chicks at necropsy. Older chickens may show upper respiratory signs (coughing, etc.) and nasal discharge ("dirty shoulders"). The very rapid onset (few days) and very high percentage of the flock that is affected (approaching 100%) is often helpful in diagnosis. Vaccines are available for administration at the hatchery by spray application or by eye drop. Birds intended for egg production or as breeders can be boosted prior to egg production onset. Treatment for secondary bacterial infections is sometimes attempted but is often unsatisfactory. Because of the highly contagious nature of this disease, biosecurity measures are usually ineffective. Vaccination is strongly recommended, especially in areas where many chickens are raised.

Laryngotracheitis: This viral respiratory disease of chickens and turkeys primarily affects the trachea, and spreads much more slowly than Infectious Bronchitis. Viral infection causes severe inflammation of the trachea and frequently causes hemorrhage. The appearance of blood at the corners of the mouth and/or nares with coughing and sneezing (causing blood in the environment) is strongly suggestive of this disease. In areas where large numbers of chickens are raised outdoors, this disease could be confused with tracheal worms. Vaccines are available for this disease but elimination through all-in, all-out production and good biosecurity is much preferable. Vaccines are only used when necessary. Antibiotic treatment for secondary infections is sometimes attempted but is often unsatisfactory.

Bordetellosis (turkeys): An important bacterial upper respiratory disease of turkeys, also known as Turkey Coryza, caused by the bacterium *Bordetella avium.* Severe coughing and other upper respiratory signs are usually noted (disease is similar to kennel cough in dogs). Disease may cause thinning of the tracheal cartilage producing an easily collapsible trachea which is responsible for death by asphyxiation. The organism is sometimes susceptible to antibiotics but treatment is often unsatisfactory. Very thorough sanitation between flocks, including water containers and feeders, is necessary to eliminate this disease between flocks in all-in, all-out production. Tends to reoccur on premises, apparently due to inadequate sanitation, persistent contamination of drinkers, and insufficient down time between flocks.

Mycoplasmosis: Several species of *Mycoplasma* may cause upper respiratory disease in chickens and turkeys (and others). Important *Mycoplasma* species cause inflammation of the upper respiratory system with accumulations of mucus with small amounts of pus in the sinuses, most often observed in the infraorbital sinus as large swellings under the eyes. The exudate associated with upper respiratory Mycoplasmosis is often much thinner than seen with Infectious Coryza (see above) which assists with diagnosis. Important species of *Mycoplasma* can also cause airsacculitis, involvement of the joints (see above), and malformation of bones in turkeys. The causative bacteria are very simple organisms, rarely resistant to common antibiotics such as tetracyclines, and do not survive in the environment for more than a few days. Although infected birds usually respond to treatment, relapse following cessation of treatment is common. Purchasing replacement stock from *Mycoplasma*-free breeders is important for control in all-in, all-out production as the Mycoplasma can be vertically transmitted to progeny from infected breeders. Because long-term flock infection is likely, and secondary infections are common, this disease has been called "Chronic Respiratory Disease or CRD".

- Generalized Respiratory Diseases

(note: Infectious Coryza, Infectious Bronchitis, and Laryngotracheitis (above) are primarily upper respiratory diseases which can also be complicated by opportunistic infections involving the lower respiratory tract)

Avian Influenza: This viral disease of poultry causes a range of signs similar to Influenza in man. Some isolates cause primarily respiratory disease, while other isolates may cause only mild respiratory disease with more severe gastrointestinal lesions and signs. Mild influenza isolates are recovered worldwide and are often of little significance. However, there are some strains that are highly pathogenic, causing intense respiratory disease and high mortality. In these highly pathogenic outbreaks, severe inflammation of the entire respiratory tract is common, with consolidation of the lungs. Occasionally intestinal lesions of proventricular hemorrhage and intestinal ulceration are observed in the more virulent outbreaks. The more virulent isolates are considered exotic and reportable in most developed countries and are controlled by vigorous testing and eradication. Vaccination is usually of limited value because field isolates may be of many different serotypes. Vaccination with the incorrect serotype is of no value. This disease may be transmitted to poultry through contact with wild birds (or their excrement), particularly waterfowl. While transmission of the highly pathogenic form of Avian Influenza to man has been documented, this phenomenon is still considered rare. Treatment of secondary infections with antibiotics may be of some value in mild cases. Local eradication is possible through strict sanitation between flocks, strict biosecurity (preventing wild bird contact), and all-in, all-out production.

Newcastle Disease: This viral respiratory disease may be indistinguishable from Avian Influenza (above). Isolates range from little disease-causing potential (avirulent) to highly pathogenic (VVND). Mild strains are more likely to cause nervous system signs than mild forms of Avian Influenza. Unlike Avian Influenza, there is only one serotype of Newcastle Disease Virus which allows for vaccination. Vaccination at the hatchery by spray or eye-drop is recommended in areas where Newcastle disease has been diagnosed. Booster vaccination of breeders or egg-production flocks prior to reproductive onset is also advised.

Mycoplasmosis: Although several species of Mycoplasma are capable of causing respiratory disease in poultry, *Mycoplasma gallisepticum* is the most common isolate associated with severe and generalized respiratory disease. Signs, treatment, and prevention are discussed above.

Fowl Cholera: This important and common bacterial respiratory disease of chickens and turkeys is caused by *Pasturella multocida.* The disease condition observed may be related to how long the infection has been ongoing. In acute (recent onset) infections, mortality may be one of the first signs. Affected birds often have dark, consolidated, and hemorrhagic lungs, filled with pus if the affected bird has survived long enough for formation. Because the causative bacteria frequently are found in the blood (septicemia) in large numbers, many organs and sites may be infected. Toxins are released from the causative organism which cause hemorrhages throughout the body, often readily observed against the lighter background of abdominal or heart fat. Birds that have died from Fowl Cholera often have severe inflammation throughout the body, particularly noticeable in the viscera, and the liver may appear to be partially cooked or "parboiled", often with multiple small abcesses. The pericardium (sac surrounding heart) is usually inflamed and may contain large amounts of pus. The survivors of acute Fowl Cholera often develop abcesses associated with the comb, wattles, or joints. The bones of the head may become infected and cause nervous system signs (known as "cholera of the head"). Isolates from commercial poultry flocks are frequently resistant to multiple antibiotics, although treatment with an antibiotic to which the causative organism is susceptible may cause marked reductions in mortality. However, many surviving birds will develop the more chronic form of Fowl Cholera following cessation of treatment. Although some commercial live vaccines are available, there are several possible serotypes of the causative and vaccination of subsequent flocks is not always recommended unless the serotype can be determined by a diagnostic laboratory (some cross-protection may be observed, depending on the isolate). An alternative to vaccination with live commercial vaccines is the use of a killed bacterin which can be produced by some diagnositic or service laboratories using the actual causative isolate. These killed bacterins can produce good control if repeated injections are made. Thorough sanitation, rodent control (and other wild animals/birds), and general biosecurity can reduce losses. It is possible to eradicate in all-in, all-out programs with significant effort applied to sanitation and future biosecurity. Several weeks of down time between flocks is strongly recommended. Fowl Cholera can easily be confused with Colibacillosis and Erysipelas and a diagnostic laboratory should be consulted if possible.

Chlamydiosis: This respiratory disease occurs in many species and ages of birds, and may cause severe and sometimes fatal disease in humans. Chlamydiosis is often confused with Fowl Cholera or Colibacillosis, and clinical signs and lesions are similar (see above). A diagnostic laboratory is necessary for differentiation. Disease may be mild but occasionally causes high mortality, especially in turkeys. This virus-like bacterium cannot be grown by normal bacteriologic methods but can be microscopically observed using special stains. Infected humans experience flu-like symptoms, often accompanied by very high fevers. The disease is easily misdiagnosed by physicians. Humans exposed to poultry or pet birds who experience flu-like disease of longer than normal duration, especially when accompanied by high fevers, should report this exposure to their physician. The disease in humans and birds is easily treated with tetracycline-type drugs, although the causative agent may be not be susceptible to some of the more common antibiotics selected by physicians for respiratory disease. Misdiagnosis in humans and failure to appropriately treat this disease can result in permanent neurological damage and sometimes death.

Avian Tuberculosis: This bacterial disease is only of concern in very old (several years) poultry. The causative agent is similar to the cause of human tuberculosis and can cause humans to react with a positive skin test. While the avian agent can cause disease in immunosuppressed humans, it is not usually a threat to normal humans. Most people become infected by exposure to wild birds/bird feces, and not through contact with poultry.

Aspergillosis: This fungal disease of poultry is caused by inhalation of large numbers of the spores. Often called "Brooder Pneumonia" because the exposure to mold spores may occur in the hatcher soon after hatching, and the resulting fungal pneumonia occurs during the brooding period during the first 1-2 weeks after hatch. Because the source of infection is environmental, this disease is actually an infectious but non-contagious disease. Turkeys or chickens in dusty environments, sometimes when raised on range, may also be exposed to infectious doses of mold spores. Occasionally, mold within housing will cause problems. Signs of this disease are coughing, sneezing, depression, etc. as expected from respiratory disease. Mortality is variable but can be significant. Necropsy will reveal clusters of nodules (usually green-yellow to gray) approximately 2-3 mm in diameter known as mycogranulomas. These grape-like clusters of spherical nodules may be found most often in the lungs and air sacs. This disease is sometimes confused with Avian Tuberculosis. Aspergillosis (Pneumomycosis) is not effectively treated with available antibiotics for poultry and treatment is useless. Nursing care (improved ventilation and maintenance of warm, comfortable temperatures, etc.) may reduce the severity and complications. Prevention is by removing the source of the mold exposure. Often physical cleaning may be sufficient, but many disinfectants are available which provide residual mold growth inhibition.

N. Diseases Affecting the Gastrointestinal Tract Only

- Esophagus-Gizzard

Fowl Pox: Occasional birds will present with lesions almost exclusively on the larynx or upper trachea. However, other birds will also show the more typical plaques in the mouth area or nodules on the skin, as discussed above, greatly assisting in diagnosis.

Crop Mycosis: An overgrowth of yeast in the crop that may cause depression, reduced feed consumption, and reduced performance, usually following antibiotic therapy for long periods of time, at high doses, or in combinations not recommended. At necropsy, the crop smells putrid or "sour", providing the basis for the synonym "Sour Crop". The normally smooth and shiny crop epithelium also becomes thickened and roughened by the yeast growth, somewhat resembling a wet bath towel in appearance. Gentian violet or copper sulfate have been used for treatment and may have some effect. However, prevention is much more effective than treatment and can be accomplished by carefully following label directions for antimicrobial use, and by limiting antimicrobial use to diseases that specifically may respond to the chosen drug.

Capillariasis: Infestation of the crop and esophagus of most birds raised for food is possible with these small round worms. While some may be up to 6 cm in length, many are smaller and appear almost like white almost hair-fine noodles at first glance. Scraping the mucosa may enhance observation of the worms. These nematodes burrow into the mucosa, causing inflammation and thickening of the mucosa. Affected birds may be anemic and emaciated. Diagnosis is through identification of the worms or their bipolar football-shaped eggs (strongly resemble whipworm eggs of dogs). Treatment with most anthelmintics is effective (pipperazine, avermectins, etc.).

Tetrameres americana: This parasitic worm (and others) may infect the proventriculus of birds. Worms are very fine ~5-20 mm in length and are usually visible. Worms may cause mucosal ulceration with hemorrhage and sometimes small circular hematomas are visible from the serosal (outside) surface of the proventriculus. These worms must have intermediate hosts for infection of poultry which may include cockroaches, grasshoppers, or pillbugs. Treatment with systemic anthelminthics is usually effective.

Gizzard/Proventricular Erosions/Dialation: Disease causing hemorrhage of the proventriculus or gizzard is most likely to occur in areas where fish meal is fed at very high concentrations. High concentrations of heated fish meal may contain toxins which directly affect the lining of the upper GI tract. Blood leaking into the upper gastrointestinal tract becomes black when in the lumen, and when birds are inverted this black liquid may passively reflux from the mouth of affected birds. This condition may be referred to as "Black Vomit". Reducing the concentration of fish meal to less than 10% of the total ration (when this is a problem) usually will result in improvement of this condition, though some fish meal may be toxic at lower concentrations. Recently, some evidence has been presented which indicates that an apparently non-related and non-hemorrhagic condition causing dilation and weakening of the proventriculus may be associated with infections bursal disease virus. Appropriate vaccination programs for gumboro have reportedly eliminated this problem (see Infectious Bursal Disease).

Fig. 69. Cycle of Infection, from Diseases of Poultry H. Van Roekel—1943.

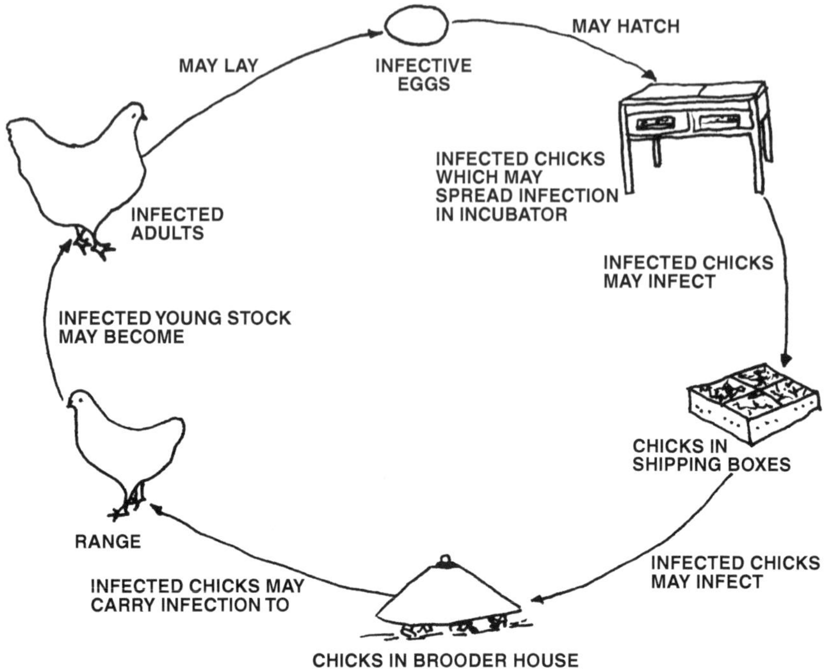

Pullorum infection in a flock may follow this cycle.

• Small Intestine

Coccidiosis: A very important and very common protozoal parasite problem of growing poultry. All of the important species for chickens and turkeys are of the genus *Eimeria.* The coccidia have a complex life cycle, allowing for tremendous amplification in the environment if birds are allowed contact with old fecal material. The oocysts are not initially infective but must sporulate in the environment. Continuous removal of fecal material or rotation of cages/pens on pasture can greatly reduce exposure of poultry to coccidia. However, oocysts are very stable in the environment and are not eliminated by down time between flocks. Each species of *Eimeria* is capable of infecting rather specific areas of the intestinal tract. Poor growth and feed conversion are common in infected flocks and passage of undigested feed is not unusual. Blood is occasionally apparent in the feces of affected birds. Necropsy may reveal inflamed and thickened regions of intestine, sometimes described as "sausage-like" due to the thickening of the intestine and the presence of reproductive stages of the parasites within the wall of the affected intestine. The reproductive stages (schizonts) sometimes appear as white streaks on the serosal (outside) surface of the intestine, and multiple pin-point hemorrhages may sometimes be observed on the inside (mucosal) surface of the affected gut. Microscopic examination of scrapings of the intestine for the presence of oocysts and reproductive stages is diagnostic. Lightweight poultry, such as egg-production strains, can be raised on wire to reduce contact with feces, greatly reducing the significance of this disease. Bamboo or wooden slats, while not eliminating the risk of disease, also reduce the significance of these parasites. Birds raised on litter with continual contact with fecal material are at high risk of development of coccidiosis. In these cases, there are 2 potential methods of control. Coccidiostat drugs are available for continual medication of feed in growing meat-type chickens. If this approach is selected, the drugs must be included continuously until just before slaughter. An alternative approach is the use of vaccines. Available commercial vaccines rely on the administration of a "controlled" exposure to low numbers of infective oocysts early in life. Immunity is generally life-long. Thus, vaccination may be the prefered approach to breeder or egg-production birds that must be raised on litter. However, it should be noted that in rural settings where birds are provided a clean and spacious environment, coccidiosis may not be a major problem even without preventative treatment or vaccination. This is particularly true if consideration is given to reducing the contact of growing poultry with fecal material by the use of wire or slats and with regular physical cleaning. Please note that wire is generally not advisable for meat type poultry because of the damage commonly inflicted on food pads and legs.

Necrotic Enteritis: This bacterial infection of the chicken intestine is caused by a primitive organism that is often susceptible to antibiotics. The disease is usually secondary to immunosuppression, filthy conditions, or both. Affected birds are often found dead or dying. Necropsy reveals distended intestines, with thin walls, containing a black, tar-like material which is the result of necrosis (rotting) of the intestinal lining and the breakdown of blood. The gut contents produce a characteristic offensive odor which as been described as "septic tank" or "anaerobic chamber" smell. In commercial production, low levels of antibiotics, such as bacitracin, may be fed continuously, with the benefit of greatly reducing the incidence of this disease. Healthy poultry, free of immunosuppressive disease and housed in clean and spacious environments, rarely develop Necrotic Enteritis.

Colibacillosis: While some strains of *E. coli* may be primary pathogens, *E. coli* often presents as an opportunistic infection with other disease problems or severe stress. Colibacillosis may present in many forms, infecting the respiratory tract (above), the navel and yolk sac (omphalitis), causing septicemic conditions very similar in appearance to Fowl Cholera, enteritis, joint infections, and etc. Enteric infections are often characterized by diarrhea, with inflammed intestines at necropsy, often with excessive mucus. Antibiotic therapy may greatly improve the performance of affected poultry, but it is advisable to select the antibiotic based on sensitivity testing as resistance is common. Generally, this disease should signal that other specific causes of stress and/or disease should be addressed as preventative measures, as this disease is uncommon in birds maintained in clean environments free of other specific pathogens.

Roundworms: Most roundworms (ascarides) are found in the small intestine of chickens and turkeys (1-4 cm) and may cause inflammation of the intestine, diarrhea, and poor performance. Intermediate hosts include grasshoppers and other insects. Very mild infections may not affect performance. Roundworms can be reduced if exposure to the intermediate host is possible, or if the intermediate host is not exposed to poultry feces. Infestations generally respond to common deworming agents, and intact worms may be passed in the feces after worming. In problem areas, regular treatment with deworming agents may be advisable.

Tapeworms: Several species of flat, segmented tapeworms may infest poultry. Surprisingly heavy infestations are often not associated with noticeable reductions in performance. Treatment is generally not cost-effective and may be associated with significant toxicity when off-label anthelmentics which are effective for tapeworms are used. Poultry tapeworms require an intermediate host which may include flies, other insects, snails, slugs, and other invertebrates. It is sometimes possible to control the intermediate host more easily than the tapeworm.

- **Ceca and Large Intestine**

Salmonellosis: There are more than 200 *Salmonella* serovars (species) which may infect domestic poultry. Almost all of these may potentially cause human infections through heavily contaminated food. Two nearly-indistinguishable serovars are host-adapted for poultry, *Salmonella gallinarum* and *Salmonella pullorum,* and cause severe disease in poultry, but are not transmissible to humans. The remaining serovars, referred to as "paratyphoid salmonellae", usually cause mild and sometimes subclinical disease in poultry, but may cause signs and lesions in poultry similar to the host adapted serovars when birds are immunosuppressed. Salmonellosis in poultry primarily affects the lower intestinal tract, causing small hemorrhages in the cecal tonsils and diarrhea frequently characterized by white pasted vents. Affected birds appear cold, huddle together, and are usually depressed. The infection may become systemic, causing petechial hemorrhages throughout the body, severe hemorrhage and infiltration of pus in the pericardial sac, and multiple lesions in the liver, frequently giving the liver a "yellow polka dotted appearance"on an unusually red background. The host-adapted serovars have been eradicated in most developed countries but the paratyphoid serovars continue to plague poultry. These serovars of lower virulence are generally not significant production problems in healthy poultry. Antibiotic therapy without specific knowledge of susceptibility is not recommended as antibiotic resistance is a tremendous problem with the salmonellae. Vaccines are generally not cross-protective against other serovars and are only useful for specific serovars. *Salmonella gallinarum* and *Salmonella enteritidis* vaccines are available in some countries where disease problems related to these serovars are endemic.

Histomoniasis (Blackhead): Protozoal disease primarily of importance for turkeys, although other poultry can be affected. Characterized by enlarged ceca filled with "cheese like" exudate and debris and peculiar lesions of the liver which are often described as "target-like" due to the concentric circles that are often evident. The causative agent, *Histomonas meleagridis,* may be transmitted by consumption of fresh feces within a flock, but initial flock infections are usually attributed to ingestion of embryonated cecal worm ova or earthworms containing cecal worm larvae within their tissues. Thus, control is often directed toward sanitation and antihistomonal drugs added to the feed (e.g. ipronidazole, nitarsone). Because histomoniasis causes much less severe disease in chickens (often unrecognized) and because cecal worms of chickens can serve as vectors of histomoniasis, it NOT advisable to raise turkeys in areas where chickens (or other gallinacious birds) are also raised. Treatment with dimetridazole or metronidazole is effective, although these drugs may not be available or labeled for poultry. Broad spectrum antibiotics may reduce morbidity and mortality as normal opportunistic enteric microflora appears to play a major role in causing the disease initiated by the causative agent of histomoniasis.

Coccidiosis: As described above, specific species of coccidia affect specific regions of the intestine. Some species preferentially infect the ceca and are somewhat more likely to cause hemorrhage in addition to the other signs described above.

O. Diseases Affecting Both the Respiratory Tract and Gastrointestinal Tract

Newcastle Disease: In addition to the respiratory disease described above, infections with highly (velogenic) forms of Newcastle Disease may also cause intestinal disease. This affinity for the intestine, in addition to the respiratory tract, is associated with velogenic strains and is referred to as "viscerotropic", resulting in the common acronym of the highly pathogenic forms of Newcastle Disease (Velogenic Viscerotropic Newcastle Disease; VVND). Intestinal erosions or ulcerations may be observed and the proventriculus may be hemorrhagic.

Avian Influenza: In addition to the respiratory disease and lesions of the internal organs described above, highly virulent forms of Avian Influenza (AI) may also cause gastrointestinal tract lesions that could be confused with VVND, although lesions caused by highly virulent AI (also called Fowl Plague) are more variable and intestinal lesions may not be present. The highly virulent forms of Newcastle Disease and Avian Influenza can sometimes be very difficult to distinguish without laboratory support.

Fowl Cholera: In addition to the respiratory disease discussed above, the causative organism of Fowl Cholera can circulate throughout the body causing localized infections in multiple sites and causing the release of large amounts of endotoxin. The intestine may be highly inflamed and may even be hemorrhagic in some cases. Diarrhea, though not a major sign, is not unusual in acute Fowl Cholera.

P. Diseases Affecting the Renal System

Ochratoxin: One of more than 200 known mycotoxins which are produced by specific molds/fungi in grain or feed prior to feeding. Production of these toxins is much more pronounced when stored feedstuffs become moist and are pronounced in areas with poor grain storage facilities, particularly in tropical or subtropical environments. In addition to reduced performance and possible diarrhea, this toxin can cause pale swollen kidneys and may be responsible for forming stones in the ureters. Renal failure resulting in visceral gout has also been reported.

Infectious Bronchitis: Some strains of this important respiratory disease (described above) also may affect the kidneys. Affected kidneys are swollen and the ureters and tubules contain large amounts of uric acid crystals, especially in young chickens. The renal tubules distended with urates have been described as appearing "worm infested" by some due to the fine white lines observed.

Q. Diseases Causing Systemic Generalized Disease

Fowl Cholera: As described above, this important respiratory disease frequently causes generalized infections following the appearance of large numbers of circulating (septicemic form) bacteria in the acute form. Hemorrhages throughout the body are caused by the toxins produced by the causative agent. These toxins may actually denature proteins much like occurs with cooking, giving the liver a "parboiled appearance". Multiple necrotic lesions throughout the liver and other organs and the eventual formation of abcesses throughout the body (joints, comb, wattles, conjuntiva, cranial bones, etc.) are hallmarks of this disease. Pus (in abcesses) in birds is often dry and hard as compared to the more liquid pus commonly produced in mammals.

Salmonellosis: In addition to the primary intestinal infections described above, *Salmonella* species can also cause severe systemic disease on occasion. This most commonly is associated with the highly virulent host-adapted salmonellae of poultry (*S. pullorum/gallinarum*) but can be related to less virulent strains in cases of stress and/or immune suppression. The systemic presentation of salmonellosis may cause hemorrhages throughout the body in acute cases and will sometimes result in the development of nodules in the heart muscle and intestine, with or without hemorrhage. The organism has been occasionally associated with chronic arthritis in birds that have survived severe initial infections. In acute infections the liver is often enlarged and red (inflamed) and may contain multiple discrete areas of necrosis, giving affected livers an appearance that have been described as "dark red with yellow/white polka dots". Control measures are described above.

Coccidiosis (bacterial leakage from intestine into blood): Although the lesions caused by coccidiosis are limited to the intestine as described above, these infections damage the intestinal mucosa (lining) and may allow large numbers of bacteria from the intestine to enter the blood stream. As the venous drainage from the small intestine goes directly to the liver, these bacteria can cause infections and problems even though they may not be specific pathogens. Liver abcesses and multifocal necrotic lesions are not uncommon in coccidiosis. Large cauliflower-like lesions sometimes appear in the liver and are usually related to E. coli "leaked" from the intestine.

Mycotoxicosis (Aflatoxicosis): There are more than 200 known mycotoxins produced by a variety of fungi/molds. Aflatoxin is the most common cause of mycotoxicosis in poultry. The toxin is produced by a specific but common species of fungus and is most often associated with corn. The fungus may actually parasitize the corn prior to harvest and pre-formed toxins may be present in field-dried corn, especially in years of drought or other plant stressors. After harvest, improper drying to less than 14% moisture or allowing moisture to rehydrate grain or mixed poultry feeds can allow mycotoxin-producing molds and fungi to grow and produce additional toxin. When adequate storage facilities for maintaining dry grain or poultry feed are not available, mold inhibiting additives are available which can be added to the mixed feed (e.g. propionic acid, gentian violet). It is important to note that these mold inhibitors do not affect levels of pre-formed toxin, they only reduce production of additional mycotoxins. Aflatoxicosis causes reduced performance of production poultry and is often associated with pale enlarged livers which may be hemorrhagic and fragile (friable). Aflatoxin also causes capillary fragility, commonly resulting in the appearance of numerous bruises due to trauma that normally would not be associated with such hemorrhage. While aflatoxicosis often contributes to the fatty pale livers associated with heavy egg production in laying hens, it should be noted that pale enlarged and fatty livers are a common occurrence in table egg producing chickens and that this can also be compounded by hen obesity. Kits are commercially available for testing feed for aflatoxin and several other common mycotoxins. Aflatoxin testing is advisable for large purchases of grain or mixed feed.

Ascites Syndrome: This pathophysiological condition has a genetic predisposition component and only affects fast growing meat type chickens (broilers). The word "ascites" refers to the development of abnormal fluid in the abdominal cavity (from any cause), hence the synonym "water belly". The specific disease known as Ascites Syndrome involves cardiac failure with resulting venous hypertension and the accumulation of transudate (essentially the cell free component of blood) in potential spaces within the body. Hydropericardium, a similar accumulation of fluid in the sac surrounding the heart, also commonly occurs. The fluid is usually clear to amber in color, translucent (not opaque), and may or may not contain gelatinous clots. Affected birds are most commonly male (2:1) and are usually the fastest growing birds within the flock. Because the condition is related to heart failure (probably involving a major pulmonary component), this condition is much more frequent and severe at higher altitudes (>2,000 M). Respiratory diseases, dust, and noxious gasses also contribute to the condition. In practice, the only preventative measure currently available is to intentionally slow the growth rate by reducing either feed intake or nutrient density of the diet. At very high altitudes it may be necessary to limit feed access of broilers from 2-6 weeks of age, sometimes to a total access time of 8-12 hours per day. However, feed restriction of this type (or skipping every 3rd day) reduces feed efficiency, increases stress, and increases the ingestion of feces and litter by the birds, thus increasing their exposure to pathogens. If possible, reducing nutrient density is therefore a preferable means of slowing growth rate.

R. Diseases Affecting the Skin and Feathers

Molt: Reproductively active poultry undergo a normal loss of feathers as they undergo seasonal or forced cessation of egg production. Affected birds may appear to be diseased at first glance but feathers will regrow prior to resumption of egg production (or reproductive activity of males). Some table egg producing strains will continue egg production well into the molt. Although severe stress or infectious disease can induce molting and cessation of reproductive activity, normal molt should not always be assumed to be disease related. Conditions of decreasing photoperiod (duration of day length) will increase the numbers of birds that undergo seasonal molt, turkeys are most photoperiod dependent. Artificial lighting is required for optimal performance when natural daylength is decreasing.

External Parasites: There are several important and common external parasites of poultry. These include lice and mites, and less commonly fleas and ticks. The most common mite affecting chickens is very small, dark in color and may be found in largest numbers in the vent area. Any confusion between mite infestations and "dirt" can be eliminated by roughly brushing the chicken over a white background. Moving "dirt" is mites. The other parasites are more easy to recognize. Permethrins (synthetic pyrethrins) are commonly safe and effective for treatment. Lice may sometimes respond to dusting with 5% carbaryl. It is important to remember that insecticides can be incorporated into eggs or meat so label directions should be followed or birds should be treated several weeks prior to egg collection or slaughter for food. Chlorinated hydrocarbon insecticides (e.g. lindane, chlordane, DDT) should be avoided as these are significant risk factors for humans consuming contaminated foods. Phosphodiesterase inhibitors are dangerous for poultry and should only be used with specific instructions or as a last resort. These compounds are not generally useful for mites and ticks.

Wear on Feathers: Caged birds sometimes develop wear patterns associated with movement against feeders, etc. This wear damage to feathers should not be confused with infectious disease.

Gangrenous Dermatitis: A condition caused by bacterial infection in immunosuppressed poultry. The disease is usually recognized in birds that are depressed, sometimes weak or ataxic, reluctant to eat, and rapidly progresses to death in individual birds. Dark moist areas of skin are observed and the causative agent may produce gas visible beneath the skin. The area under the skin may contain a reddish fluid and may be swollen. Numerous antibiotics seem to have sometimes reduced the severity of outbreaks but frequently fail. Prevention is most appropriately aimed at the underlying immunosuppressive disorder(s) that predispose to infection with the opportunistic organisms involved in this condition.

Cannibalism, Scratches, and Scabs: Fighting and aggression are sometimes a significant problem for poultry. Aggressiveness is more of a problem with males than females but can be significant for either sex. Aggressiveness increases with the onset of puberty and sexual maturity. Effective management practices including beak dubbing (approximately 1/3 from tip of beak to nares) and reduction of light intensity if possible. The addition of brightly colored soda cans (particularly red) appear to sometimes reduce cannibalism problems.

Tumor causing Diseases: There are several cancerous tumors that involve the skin of poultry. However, the principle such disease with sufficient frequency to be a recognizable problem for small flocks is Marek's disease of chickens. As described above, the causative virus causes feather follicle tumors and these red, inflamed and swollen tracks of follicles are helpful in diagnosing this disease. These lesions are most often observed in plucked carcasses. There is absolutely no suggestion of any detrimental effect of this virus and human health and many (if not most) normal appearing chickens may be infected with this or similar viruses. The meat from affected chickens is therefore quite edible, although condemned in the U.S. for aesthetic reasons. Control is described above. A second less frequent problem involves a leukosis family retrovirus called J-virus.

Section 8

Marketing

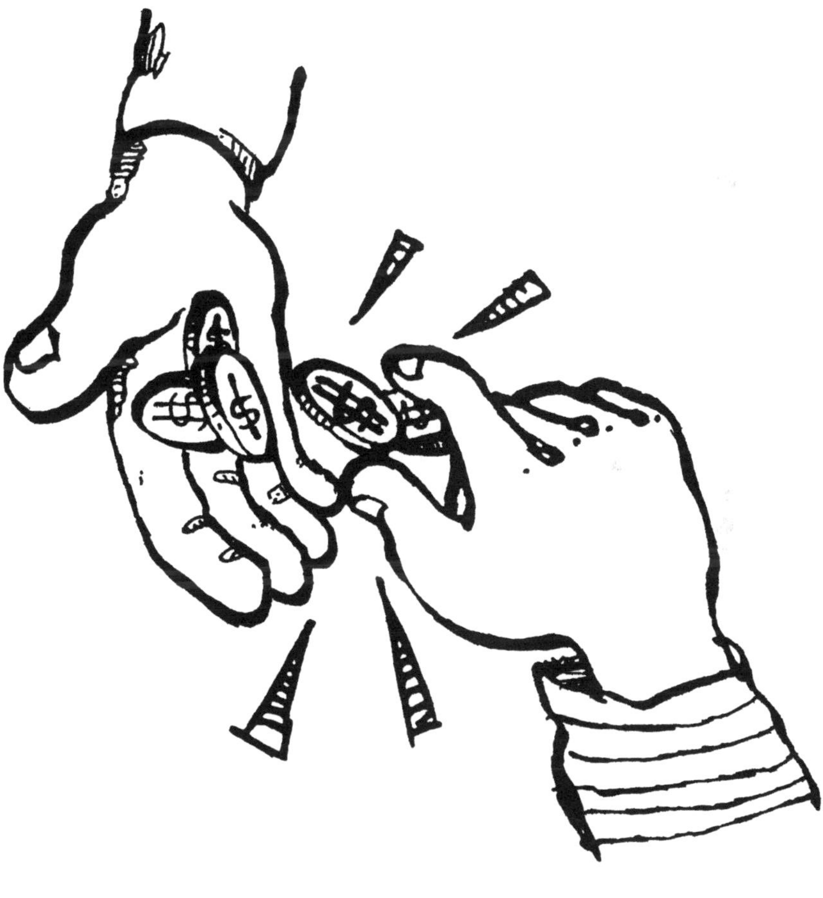

A thorough study of the potential poultry market must precede any plans to expand production. Both eggs and meat are perishable commodities so storage and transportation to market are major factors in success. Financial failure often occurs if prices drop or the method of transporting products to market is cut off.

A. Eggs. Egg quality diminishes rapidly at higher temperatures or if left in the sun. In hot weather some producers gather eggs as frequently as 5 times a day and move them to cool storage. Ideal storage temperature is 4 to 13°C (40 to 50°F). Since refrigerators are seldom available under primitive conditions, temporary cool storage is sometimes improvised by using a cave or burying eggs in a clay pot in a shaded area. They may be placed on straw or a mat and covered with a damp cloth or straw. Soil around the pot should be moist without standing water.

Three other methods of storing eggs for home consumption may prolong their keeping qualities. 1) Oil in the form of a thin film will prolong quality for three weeks or longer if stored under 10°C. Eggs should be dipped in a light mineral or cooking oil such as coconut oil. The oil temperature should be about 11°C higher than room temperature. If reused, the oil should be filtered and sterilized by heating to 116°C. 2) Water glass. Eggs are maintained in solution of one part water glass (sodium silicate) to 5 parts of previously boiled but cooled water. Eggs keep for several months if covered and stored in a cool place. 3) Lime water solution is made by mixing 2.3 kg of finely powdered quick lime with 6 liters of boiled but cooled water. After the solution has settled overnight eggs may be stored for more than a month.

Soiled eggs may be cleaned by rubbing them lightly with fine sandpaper, emery cloth or steel wool. Washing is generally discouraged as it may cause bacterial contaminants to be drawn inside the egg through the shell. Since eggs vary greatly in size, they are sometimes sold by weight. In the United States they are graded by weight into 5 sizes: small 35-42 grams, medium 43-49 g, large 50-56 g, extra large 57-63 g and jumbo 65 g and over. Defective eggs may be detected by candling to remove eggs with cracks, blood spots, mold, rot, or parasites using candling methods (Section V). Attractive packaging in cardboard or plastic containers serves for protection and customer appeal.

B. Meat. Carcass quality deteriorates so rapidly at room temperature that chickens are frequently sold live and dressed by the customer. Although cooking kills all harmful bacteria in chicken meat, “food poisoning” due to bacterial growth may occur if there is extended storage of either cooked or raw poultry meat at room temperature. Refrigeration methods of preserving dressed poultry has revolutionized the poultry industry. With assembly-line procedures, elaborate processing plants use mass production methods of dressing, storage, and shipment.

Broilers may be shipped on ice for thousands of miles by refrigerated trucks. If frozen they may be held for several months and shipped by boat around the world. Development of these large scale methods has resulted in great reductions in the price and availability of poultry meat. Since the risk of spoilage is reduced by shipping freshly frozen birds, this method is preferred in some countries.

C. Financing. Enlarging a poultry enterprise usually requires financial assistance in the form of short-term or long-term loans. Small poultry farmers generally have had good records for loan repayments. Possible funding sources would include:

a. International agencies. The World Bank, Oxfam, U.S./A.I.D., Grameen Foundation, UNICEF and others agencies sponsor rural development programs with funding assistance.

b. Others. Cooperatives, banks, credit unions, women's groups, chick suppliers, feed manufacturers, integrated poultry companies, and local money lenders may be good credit sources for capital funds required in building facilities. They may also finance short term loans to purchase chicks and feed.

c. Government agencies. In the USA federal and state agricultural loan programs have been established through administrative or agricultural extension networks. Secondary school (4H or FFA in the USA) poultry projects have often initiated young people into careers with the poultry industry.

D. Contracts. Lenders may work out purchase agreements with growers. Drawing up specific management and repayment schedules may help to overcome lender reluctance. Details may include: records of past experience, a current balance sheet, projected income statement, chronological management plan, schedule for visits by advisory personnel, and a marketing and repayment plan.

Section 9

The Advantages of Raising Ducks

Mallard Drake

Mallard Duck

Pekin Drake

Pekin Duck

Rouen Drake

Rouen Duck

From ancient times domestic ducks have served as a source of food and income for people in many parts of the world. Ducks are a source of meat, eggs and down-feathers (for making bedding and warm jackets). Ducks are able to subsist and grow to maturity on relatively simple diets, based on locally available feedstuffs. Duck meat and duck eggs are good dietary sources of high quality protein, energy and several vitamins and minerals. When properly included as part of a well balanced daily diet, duck meat and eggs can supply a substantial portion of the nutrients required by humans. Ducks may be raised in small or large flocks. A small flock of ducks may be kept by a household as a supplemental source of food or income. A small flock of ducks can be established at low cost. A higher investment is required to establish larger, or commercial flocks, which require better buildings, equipment and feeds. However, greater income, supporting several families may be realized if a large flock is properly managed.

Fig. 70. Young Pekin ducks.

A. Duck Breeds

Domestic ducks fall into the following major genetic classifications.

(1) Common ducks. Most domestic ducks fall into this group. Common ducks are believed to have originated from the Mallard (Anas platyrhynchos). Some of the better known breeds of common ducks include the Pekin, Asylesbury, Rouen, Call, Indian Runner, Khaki Campbell, Cayuga, Albio, Maya, and Tsaiya. Common ducks can interbreed, and produce fertile offspring. Eggs from common ducks require about 28 days to hatch.

(2) Muscovy ducks. The Muscovy (Cairina moschata) is unrelated to common ducks. This breed is believed to have originated in South America, although ancient records of this or a similar breed have been found in Egypt. There are both colored and white feathered varieties of Muscovies. The Sudani is a breed of Muscovy found in Egypt. Unlike common ducks, the head and face of Muscovies is covered with caruncles (a fleshy growth that resembles wattles). Another prominent feature of Muscovy ducks is the large difference in body size between the drake and the duck, the male weighing 30-50% more than the female. Muscovies tolerate hot weather much better than common ducks. Muscovy eggs require about 35 days to

Fig. 71. Muscovy drake.

Muscovy leaving nest is frequently due to mites or lice.

3:1 feed conversion in ducks.

hatch. While Muscovies can be crossed with common ducks, the offspring is sterile and therefore cannot reproduce.

(3) Sterile Hybrid Ducks. When Muscovies and common ducks are allowed to mate naturally, the fertility rate is usually very low. It is a common practice today to use artificial insemination to increase the fertility rate when these two species are mated. Whatever the method of mating, the offspring are sterile and cannot be used for breeding. These hybrids are usually raised for their meat, or in some cases for their liver (foie gras), which is sold to restaurants. These sterile hybrids are called mule or hinny ducks. In some cases special names are assigned to hybrids by commercial breeders. For example, one hybrid produced by crossing Muscovy males with Pekin females is called "Moulard". Such names may identify the commercial breeder and the particular strain of Muscovy and common duck used to produce the hybrid. In Taiwan, the hybrid produced by crossing a White Muscovy male with a Kaiya (Pekin x Tsaiya) female is called simply, the "Mule Duck". Mule Ducks (Figure 72) are popular among the people of Taiwan because of their taste and high proportion of lean meat.

Fig. 72. Taiwan mule ducks.

B. Choosing The Right Breed of Duck

Choose a breed of duck that best suits your needs. This may be a native breed that is well adapted to the weather conditions of the area where you live, or it may be a breed that performs better than local breeds, available from duck keepers in other areas or from a commercial breeder.

(1) Meat Type Breeds

If you are mainly interested in meat, and you are able to obtain the necessary feed or feed ingredients required, choose a meat-type breed like the Pekin (see Figures 70 and 87). Pekin ducks grow rapidly, reaching approximately 90% of their adult weight at 7 weeks of age, when properly fed. It is not uncommon for commercially grown Pekin ducks to weigh 3.2 kg (7 lb.) at 7 weeks of age. Their growth rate under less favorable conditions will depend upon the quality of the diet they are fed. But even under less optimum conditions, Pekins can do quite well. The meat from Pekin ducks is very tender and juicy and is known worldwide for its delicious taste. Some meat-type breeds, such as the Aylesbury and Rouen, and unimproved lines of Pekins as well, have become less popular in recent times due to improved commercial lines of Pekins or other white-feathered Pekin-like ducks, such as Cherry Valley (England), Maple Leaf (USA), Legarth (Denmark) and Steggles (Australia).

The Muscovy duck is also a popular meat duck in some areas of the world. This breed is presently very popular in France. The large breast muscle of the male (often weighing 700 g at 13 weeks) gives this breed an advantage over common ducks which have considerably less breast muscle. However the large difference in body size between the Muscovy drake and duck , as already mentioned, as well as the fact that Muscovies have lower fertility rates than most breeds of common ducks, are major shortcomings. However, commercial Muscovy breeders in France have partially overcome this reproductive deficiency through genetic selection. As a meat duck, mule ducks are an improvement over the pure Muscovy in that the difference in size between males and females is not nearly as great, and it is superior to the Pekin in that it has more lean meat.

(2) Ducks for Egg Production

Khaki Campbell duck

If eggs for human consumption are the product desired, choose a high egg producing breed of duck such as the Khaki Campbell (Figure 73), the Tsaiya or Indian Runner. These breeds are capable of laying in excess of 230, and in some cases, over 300 eggs per year. These breeds are usually considerably smaller in body size than meat-type ducks. Although Pekin ducks are usually bred for their meat, some high egg producing strains of Pekins have been developed. Cherry Valley Farms in England as well as other commercial duck breeders have developed a number of lines of high egg producing Pekin-type ducks.

Fig. 73. Khaki Campbell duck.

(3) Multi-purpose breeds

Often ducks are kept as a source of both meat and eggs, and for their feathers as well. A number of breeds such as the Pekin, Aylesbury, and Maya (China) can meet these requirements. Some compromise is necessary when a strain is selected for both meat and eggs. Generally ducks that reach heavy weights at market age are not the best egg layers, and ducks that are good egg producers, are smaller in body size. A dual purpose line is usually selected for moderate body size and satisfactory egg production. Local breeds are often good choices if both meat and eggs are needed.

(4) Ducks for Herding

For centuries, ducks in the rice producing areas of the Orient have been managed under the traditional herding system (Figure 74). Under this system, native ducks are selected for generations for their ability to glean most of their food from harvested rice fields, levees, swamps and waterways. A flock of herded ducks may be the major source of income for one or more families. Examples of breeds of ducks selected for herding are the Alabio and Bali of Indonesia and the native Maya in China. The name "Maya" (house duck) refers to a very common duck found in the rice growing areas of China. The feather coloring of Mayas resembles that of the female Mallard, and for this reason, Mayas are sometimes called "Sparrow Ducks". In addition to the Maya, there are a number of distinct lines or breeds in China, such as the Gaoyou, Baisha, Yellow Colophony, and of course the Pekin (Beijing duck).

Fig. 74. Herded ducks

For suggestions on securing breeding stock, see Section 1 of this book.

Typical growth rates of three representative breeds of ducks are shown in Table 9.1.

C. Duck Housing and Management

Ducks adapt well to a wide range of management systems provided they receive essential basic care. Except for the early brooding stage, when ducklings require a higher temperature and special attention by the caretaker, which is discussed in more detail later, their basic needs are as follows.

(1) Protection from predators and extremes in weather.

(2) A clean, dry sheltered area. Although ducks can spend most of their time outdoors, on ponds or in wet areas, they require a clean dry sheltered area where they can retreat, rest, clean and preen their feathers. This allows the ducks to waterproof their plumage which protects their skin from injury and helps keep their body warm.

TABLE 9.1 Typical Growth of Representative Breeds of Ducks

	Pekin		*Muscovy*		*Khaki Campbell*	
	Male	*Female*	*Male*	*Female*	*Male*	*Female*
Age Weeks	*Body Weight in Grams*		*Body Weight in Grams*		*Body Weight in Grams*	
0	55	55	45	45	44	43
1	270	290	125	120	150	140
2	800	810	290	270	400	350
3	1400	1380	600	540	670	550
4	1925	1930	1000	830	940	750
5	2460	2450	1500	1160	1050	910
6	2950	2845	2060	1530	1210	1080
7	3280	3110	2600	1900	1290	1180
8	—	—	3130	2130	1380	1210
9	—	—	3550	2250	1420	1240
10	—	—	3800	2350	1510	1270
12	—	—	4300	—	1740	1420
16	—	—	—	—	1750	1470
18	—	—	—	—	1770	1570

(3) Clean water for drinking (water that is free of germs and toxins harmful to ducks). Water for swimming is not essential, but can be beneficial in areas where temperatures are high.

(4) A diet that provides all of the duck's daily nutritional needs.

(5) Adequate light stimulation, especially for layers.

(6) Protection from disease.

The small home flock. A small flock of ducks can be kept by a household at a low cost. Except for a brooder, which is needed for the first week or so, the main things needed to get started are yard space, a simple structure, such as a shed, inexpensive fencing, a feed hopper or trough made of wood and a simply constructed watering device (see Section 4). The shelter should be located on a high, well-drained area of the yard so that water will drain away from the area when it rains. Whenever available, a sandy soil is preferable for the duck yard because it drains quickly after a rain. The earth floor of the sheltered area should be bedded with straw, shavings or similarly dry absorbent material. Low fencing (about 61cm) is satisfactory for Pekins, since they do not fly, but not for Muscovies, which are adept to becoming airborne. If predators are a problem at night, the open areas of the shed may have to be covered with an inexpensive netting or wire mesh.

Fig. 75. Duck/Fish pond

Raising ducks on open ponds. Ducks may be kept successfully on open ponds, provided a nearby dry sheltered area is available. Ducks kept on ponds may obtain part of their food from plant and animal life in and around the pond, but supplemental feeding will probably be necessary. In tropical areas it is common to combine duck raising on ponds with fish farming (figure 75). Ponds are stocked with fish such as Tilapia which are raised for human food. Manure from the ducks provides nutrients for growth of plankton which the fish consume. The number of ducks kept on ponds must be limited to prevent an over-supply of nutrients and overgrowth of plant life which will cause depletion of oxygen in the water and kill the fish. Usually both the ducks and fish are given supplemental feed, which on commercial duck/fish farms is often a nutritionally complete pelleted ration.

Herding duck management. In Indonesia, herded flocks under the care of a single herdsman usually range in size from 90 to 130. During the day, a flock of ducks, usually mature females, is allowed to search for food in areas where food is most plentiful. At night, the flock is returned to a confinement, usually a bamboo pen, where eggs are laid during the night. Eggs are collected and sold, or used for food by the herdsman's family. The major part of the diet of herded ducks consists of whole grains and snails, plus small amounts of insects, leaf material, crabs and frogs. It is the job of the herdsman to move the flock, as often as necessary, to areas where food is plentiful. Portable fencing and other equipment is moved with the flock to each new location. A grassy area with some protection, such as provided by trees, is selected as a base camp where the fencing is set up. Supplemental feed is given to herded ducks only when the food supply in the fields is inadequate.

Commercial duck production. Since this book is intended mainly for people who will raise poultry in small flocks, commercial duck production will not be discussed in any detail (Figure 76). For those who are interested in commercial duck growing, some sources of information are listed in Section 11.

Fig. 76. Modern duck building

Brooding. Much of the information on brooding chicks, presented in Section 5 of this book, can be applied to ducklings. If ducklings are hatched artificially, rather than by a broody duck, the caretaker must provide the newly hatched ducklings with a warm dry brooding area free of drafts, with a source of heat, such as those described for chicks in Section 5, and feed and drinking water located near the heat source so that the ducklings learn to drink and eat soon after they are placed in the brooder. If ducklings haven't learned to drink within a few hours, it may be necessary to dip their bills in the drinking water in order to coax them to start drinking. The floor of the brooding area should be bedded with clean dry litter such as wood shavings or chopped straw. Make sure the litter is free of mold. Sometimes newspapers are put on the floor for the first few days. Papers must be changed as often as necessary to kept the area clean. Use brooder guards to keep the ducklings confined to the area where the heat, water and feed are located (see Section 5). The brooder guards should allow enough room so that the ducklings can move away from the heat if it gets too warm. See Table 9.2 for recommended temperatures, which are gradually lowered as the ducklings grow. In addition, ducklings should be allowed access to more of the floor area of the pen as they grow older. When outside temperatures are above 21.1°C or 70°F, ducklings can be allowed outdoors part of the day after about 14 days of age.

Optimum temperatures for ducks. At the time of hatching, ducklings require a high temperature of about 29.4°C (85°F). They are not yet able to regulate their body temperature and must have supplemental heat such as that provided by a brooder. As they grow older they become better able to produce and conserve heat, and regulate their body temperature. After a duckling is fully covered with feathers and down, they are able to maintain the proper body temperature even when the outside temperature is low. The recommended temperatures for ducks at different ages are given in Table 9.2.

TABLE 9.2 Optimum Temperatures for Ducks

Age in Days	*°F*	*°C*
1	86	30
7	81	27
14	73	23
21	66	19
28	59	15
35	55	13
42	55	13
49	55	13
Developing breeders	55	13
Laying breeders	55	13

TABLE 9.3 Floor Space Allowances For Ducks [1]

Age in Weeks	*Space per duck (cm²)*	*Space per duck (ft²)*	*Dimensions square area 10 ducks (m)*	*Dimensions square area 100 ducks (m)*
1	289	0.31	0.54 x 0.54	1.7 x 1.7
2	576	0.62	0.76 x 0.76	2.4 x 2.4
3	1024	1.10	1.01 x 1.01	3.2 x 3.2
4	1369	1.47	1.17 x 1.17	3.7 x 3.7
5	1764	1.90	1.33 x 1.33	4.2 x 4.2
6	2116	2.28	1.45 x 1.45	4.6 x 4.6
7	2304	2.48	1.52 x 1.52	4.8 x 4.8
Developing Breeders	2500	2.69	1.58 x 1.58	5.0 x 5.0
Laying Breeders	2809	3.02	1.68 x 1.68	5.3 x 5.3

[1] Floor space inside shelter, allow at least twice this amount outdoors.

Floor space. Overcrowding ducks can be extremely detrimental to their health, growth or egg production. Providing adequate floorspace at each stage of development is basic to successful duck raising. It is better to stock ducks at approximately the recommended density (see Table 9.3) in cold weather so that body heat will help warm the room in which the ducks are confined.

Flooring for ducks. Duck keepers should take care to avoid flooring that will injure the skin covering the feet and hock joints of ducks. The smooth skin of ducks is not as tough (not as cornified) as that of land fowl, and is more susceptible to injury when ducks are confined on surfaces that are too hard, rough, or abrasive. Slats, wire floors or cages may cause substantial injury to the feet and legs of ducks, depending upon how smooth and non-abrasive, and free of sharp edges these surfaces are. Stones, mixed with the soil covering the duck yards can also cause injury. The detrimental effect of flooring on ducks increases with the age and size of the duck, and the longer ducks are confined to the flooring. If ducklings are kept on wire floors for only the first week or two and then moved to quarters with litter floors, foot and leg injuries will be minimal, unless the wire is very

Fig. 77. Wire flooring

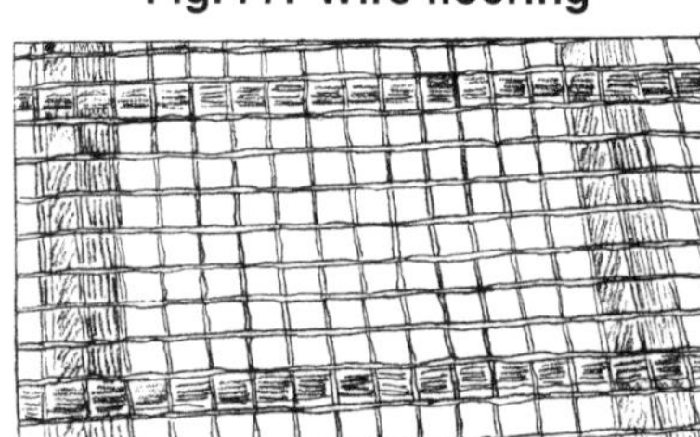

poorly constructed. Injury is also greatly reduced if wire occupies no more than from one-fourth to one-third of the floor area. Large breeds, like the Pekin are more likely to sustain injury from wire than smaller breeds like the Khaki Campbell. Properly constructed wire floors are usually a better choice than slats, which sometimes cause leg deformities as well as injury to skin. If wire floors are used, floors for ducklings under 3 weeks should be constructed of 1.9 cm (3/4 inch) mesh, 12-gauge welded wire, attached to a frame designed to keep the wire flat, and minimize manure accumulation (see Figure 77). For ducks over 3 weeks, 2.5 cm (1 inch) mesh is best. Vinyl coated wire is preferable, but smooth galvanized wire is satisfactory. It should be kept in mind that clean dry litter-covered floors (earth or cement) and clean sand covered yards are probably the least injurious surfaces of all for ducks.

Management of litter and yards. Ducks drink and excrete more water than chickens. Their droppings contain over 90% water. It is therefore necessary to take extra measures to maintain litter floors inside sheltered areas in a dry condition. This will require regular addition of fresh bedding, on top of the bedding that has become soiled or wet, and when necessary, cleaning out the old litter and replacing it with fresh litter. After ducklings are old enough to spend most of their time outdoors during the day (usually after 3 weeks), waterers should be located outside, as far away from the house as possible. This will reduce tracking water to the litter. In area where temperatures drop below freezing, water will have to be provided indoors, when necessary. Duck yards should be maintained in a clean condition by removing the upper few inches of soil and replacing it with clean soil (preferably sand) whenever necessary.

Fig. 78. Feed hopper

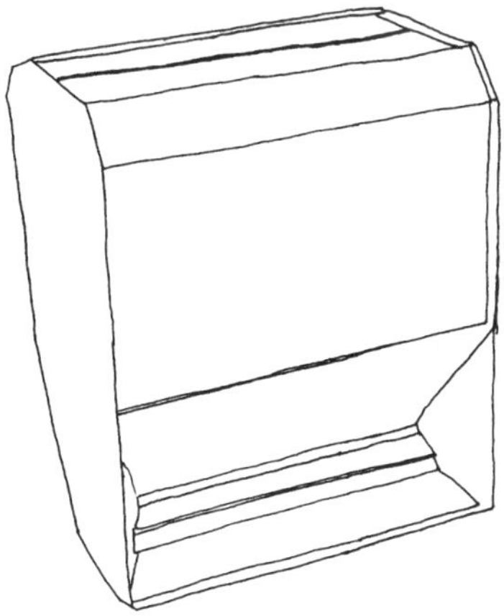

Feeders and feeding space. Most feeders used for other poultry, are satisfactory for ducks, provided differences in eating habits are taken into account. Ducks eat with more of a "shoveling" or "scooping" motion than with a pecking motion as do chickens. If ducks are hand fed, simple trough feeders work fine. If feed hoppers are used they should be constructed so that feed will slide down in the hopper as feed is consumed at the bottom, and sufficient space is allowed in the feeding area for ducks to exercise their shoveling eating motion. Providing an apron in front of the feeding area, for catching feed that is dropped or billed out, will reduce feed wastage (see Figure 78). During the early stages of growth, ducklings eat frequently, much like chickens. As they grow older they are able to store increasing amounts of feed in their esophagus (gullet) at each feeding, and thus need to eat less frequently. By about four weeks of age, Pekin ducks can easily consume 100 grams or more of pellets at a single feeding. It is important to provide about 2.5 cm (1 inch) feeder space per duck for about the first 3 weeks. Afterwards this can be gradually reduced to about half this amount so long as there is no crowding at the feed hoppers. Developing breeders on restricted feeding should be allowed plenty of feeding space so that all birds can eat at once, which requires about 4 inches (10 cm) of linear space per duck.

Waterers for ducks. Waterers that are satisfactory for chickens and turkeys are usually acceptable for ducks, as long as the size of the duck's bill is considered. Trough, can or jar-type waterers can be used so long as the drinking area is wide enough (at least 4 cm for older ducks) for the duck to submerge its bill. The same requirement applies to automatic trough, cup or bell type waterers (see Figure 79). Nipple waterers (see Figure 80), if properly adjusted for the duck's height, are also satisfactory. If waterers are located indoors where the floor is bedded with litter, waterers should be located on a wire-mesh screen to reduce wetting of the litter (see Section 4). In commercial duck houses, it is usually advisable to construct a cement floor drain underneath the water screens. For starting and growing ducks, provide a minimum of about 1 inch (2.5 cm) of linear watering space per duck. Increase this to 2 inches (5.0 cm) per duck for developing and laying breeders. If nipple waterers are used, provide 15 nipples per 100 ducks for starting and growing ducks and 20 nipples/100 ducks for developing and laying breeders. Starting ducklings should always have access to watering cans, jars or troughs until they have learned to drink from nipple waterers.

Fig. 79. Bell-shaped waterer

Fig. 80. Nipple drinkers

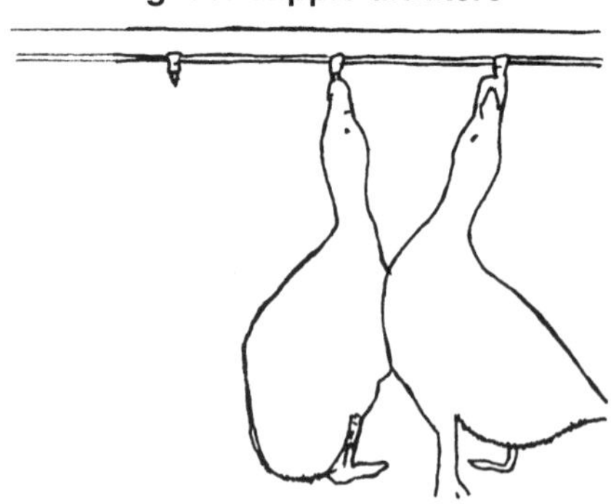

Ventilation. Duck houses or shelters for small flocks usually do not require mechanical ventilation as used in modern commercial duck buildings. However some ventilation is always necessary when ducks are kept in a house enclosed on all sides. Window openings, and ridge ventilation may provide adequate air exchange. If flocks of over 100 ducks are kept in totally enclosed houses, the use of ventilation fans may be necessary. See Section 11 for a list of books that discuss ventilation of commercial duck buildings.

Lighting. The length of the laying period of ducks can be increased considerably if supplemental lighting is provided. If supplemental light is not provided, egg production will be seasonal and dependent on changes in natural daylength. Adding artificial light to extend the daily light period to about 14 hours will provide adequate light stimulation for ducks to lay continuously for 7-12 months. Prevent any decrease in day length. If ducks are allowed outdoors and get plenty of natural light during the day, the usual practice is to turn artificial lights on at a set time before sunrise, off at a set time after sunrise, then on again before sunset and off after sunset, maintaining a constant light period of 14 hours (and a dark period of 10 hours) each day. Such a lighting schedule is usually accomplished with the aid of electric time clocks that turn lights on and off at set times. If timers cannot be used, lights can simply be left on 24 hours a day, or turned on manually before dark and off after daylight. Forty watt light bulbs, 2.4 m (8 feet) to 3 m (10 feet) high, spaced 4.3 m (14 feet) apart will provide sufficient light (one foot candle at duck eye level) to stimulate egg production. Artificial lighting is less important for growing ducks. Ducks are nocturnal, and can find feed and water in the dark. However artificial light is important the first few days to assist ducklings in getting started drinking and eating, and it is recommended that some supplemental light be provided for at least the first 3 weeks. Totally confined growing ducks, as in commercial production, will require some artificial light every day. It is also beneficial to provide dim light (use low wattage bulb) at night to help prevent stampeding if the flock is disturbed.

Preventing cannibalism in ducks. Although ducks may be somewhat less cannibalistic than chickens, they do sometimes pull out feathers causing bleeding and the development of pinfeathers that are hard to remove when the duck is slaughtered. If this problem develops, shortening the upper bill, by trimming or searing the tip to retard its growth, and thereby preventing the duck from grasping feathers, may be necessary. Trimming the upper bill can be accomplished with scissors or with an electric debeaker, such as one made by Lyon Electric Company (address is given in Section 5 of this book). Bill Heat Treatment (BHT) of the upper bill, to retard its growth, is also done with an electric debeaker. At day of hatching, the upper bill is brought in contact briefly with a moderately hot stationary blade, bent at a 45° angle, searing the area of the nail (see Figures 81 and 82).

Fig. 81. Bill heat treatment

Fig. 82. BHT blade

Management and feeding of developing breeders. If ducks are to be kept as breeders, they should be kept separate from market ducks. Developing breeders should not be fed all the feed they will eat (of concentrate feeds) and thus be allowed to become excessively fat. Restricting their daily feed intake as they develop will: (1) delay sexual maturity, (2) increase egg production, (3) increase egg size, (4) reduce mortality, and (5) increase fertility and hatchability of eggs. Feed restriction should be begun as soon as possible after the starting period (first 2-3 weeks). However for practical purposes, it is usually begun at around 7 weeks when potential breeders are selected from the flock. Full instructions for limited feeding are given later in this section. Artificial lights should not be turned on in pens where developing breeder ducks are kept, since light stimulation, above that provided by natural light, may incite breeders to start to lay too early. It is desirable to delay the start of lay in Pekin ducks to 28 weeks. This is accomplished by restricted feeding and avoiding extra light stimulation.

Management and feeding laying breeders. About three weeks before breeders have reached laying age, they should be housed at the location where they are to stay throughout their laying cycle. The house should be equipped with lights, nests and feeders. If breeder are allowed outside during the day, which is a common practice except in cold climates and on large commercial farms, waterers should be located outdoors away from the house. The breeders should be closed inside the laying house at night, and the feeders should be closed, as well. This treatment allows the breeders time to adjust to the house, equipment and routine before they begin lay. Once they begin laying, they are much more likely to lay in the nest. Artificial lights should be turned on about three weeks prior to the beginning of lay, as instructed under "lighting" in this section. Since drakes may lag behind ducks in sexual development and are sometime slow in learning to mate, fertility rate will increase sooner if males are about a month older than females. About 2 weeks before laying begins, switch breeders from developer to breeder feed and gradually increase their daily feed allotment so that by the time they reach peak egg production they will be on full feed.

Ratio of drakes to ducks. In meat-type ducks such as Pekins, a ratio of one drake to five females is usually optimum. If there is excessive injury of females by drakes, the number of drakes should be reduced to 1 drake/6 ducks. In egg laying breeds of ducks, such as Khaki Campbells and Tsaiyas, male to female ratios of 1:6 to 1:7 will often give optimum results.

Nests for breeders. Since domestic ducks habitually lay their eggs at ground level, nests for ducks should be located at floor level inside the breeder house (see Figure 83). Nests are usually located around the perimeter of a pen, next to the inside walls so that they do not interfere with movement of the flock inside the pen. Provide 1 nest for every 3-4 females. Nests should be bedded with clean dry litter such as straw or shavings. Bedding should be added as often as necessary to keep the eggs clean. When the nests become too full of bedding or the bedding becomes wet or soiled with duck manure, it must be removed and replaced with clean litter. Many, if not most of the problems of hatching duck eggs, come about as a result of eggs becoming soiled with duck manure under wet conditions, allowing harmful bacteria to enter the egg, multiply and infect the developing embryo. In addition many of the health problems of ducklings during early life are associated with dirty eggs. Preventing this contamination, by keeping the nest and the

Fig. 83. Duck nest

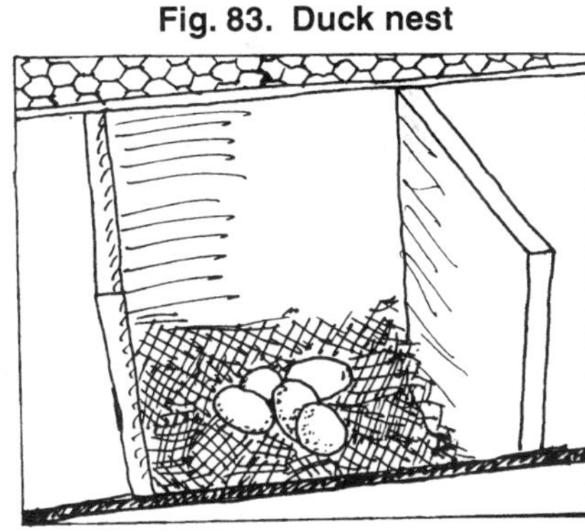

breeder pen litter clean and dry, will go a long way toward assuring good hatchability and healthy ducklings.

Egg Collection. Since ducks lay most of their eggs early in the morning, before 7 a.m., eggs should be collected around 7 a.m. and at least once again later in the morning around 10 a.m. Since ducks may bury their eggs in the nest, egg collectors should dig down in the bedding with their hands to make sure they have collected all the eggs. Some eggs may be laid outside the nest on the litter, or in the yard. Yard eggs should not be set, since they are likely to be contaminated with harmful bacteria. These eggs can be used for food if they are used soon and thoroughly cooked.

Fig. 84. Pekin duck eggs

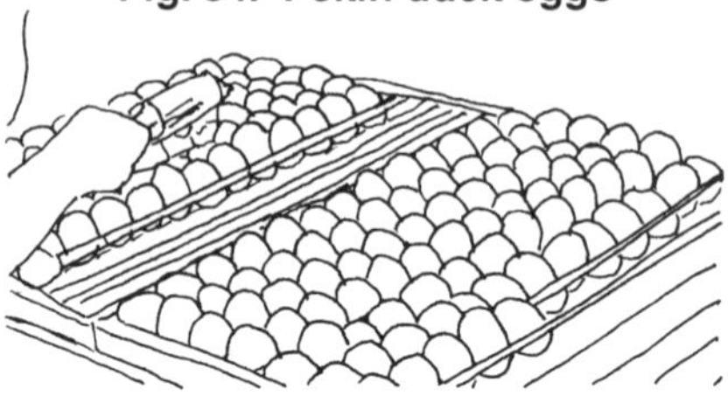

Forced molting of breeders. After breeder ducks have been laying for 7-12 months and egg production has dropped below about 40%, it may be beneficial to induce the breeders to stop laying completely and go through a molt. After the molt, egg numbers should increase and shell and interior egg quality should be improved. It will require 8-12 weeks for breeders to go through a molt and start laying again. To induce breeders to go into a molt, it is important to stop all breeders from laying as soon as possible. The steps to follow in bringing on a molt and bringing breeders back into lay are listed below.

(1) Stop all supplemental lighting (do not turn on artificial lights day or night). Do not allow breeders access to feed for 3 days, or until all egg laying ceases. Allow drinking water, but no feed.

(2) Once egg production has stopped, start feeding a restricted daily amount of a duck maintenance or developer ration (follow instructions for restricted feeding of developing breeders, given elsewhere in this section).

(3) After about six to eight weeks, or about 3 weeks before laying is to resume, return breeders to their regular lighting program of 14 hours per day (see section on lighting).

(4) About 2 weeks before laying resumes, start gradually increasing their daily feed allotment, and switch from developer to a breeder-layer ration. By the time the breeders reach peak egg production, they should be on full feed.

D. Hatching Duck Eggs

Most of the information on incubation and hatching, given in Section 5 of this book, can be applied to ducks. Since duck eggs are larger than chicken eggs, setting trays must be designed to accommodate their larger size. Eggs from common ducks like Pekins require 28 days to hatch. Eggs from Muscovy ducks hatch in about 35 days after setting. When larger numbers of duck eggs are to be hatched, large commercial incubators (setters) and hatchers are normally used. Pekin duck eggs are kept in a setter for 25 days and then transferred on the 25th day to a hatcher where they remain until they hatch on the 28th day. Eggs are automatically turned while in most setters (usually hourly). Turning is not necessary in the hatcher. Basic procedures and conditions for hatching duck eggs are as follows.

(1) A day or two before setting eggs, start the incubator and allow the temperature and humidity to stabilize. Set the temperature at 37.5°C (99.5°F) and relative humidity at 55% (29.2°C or 84.5°F on wet bulb thermometer). Set ventilation as recommended by the incubator manufacturer, or if the incubator is homemade, see Section 5. Eggs must be turned, either mechanically, or by hand a minimum of 4 times a day, and preferably hourly.

(2) Select eggs to be set by careful inspection and candling. Do not set eggs that are cracked, double yolked, misshapen, oversized, undersized or dirty. For best results, set eggs within 1-3 days from the time they were laid. There is an average loss of about 3% hatchability for eggs stored 7 days before setting, and about 10% loss for those stored 14 days. Except in the case of small incubators that have no trays, set eggs small end down. If eggs have been stored in a cooler, take them out of the cooler the night before setting and allow them to come to room temperature.

(3) On the day of setting, put eggs in incubator, close doors and allow incubator to come to operating temperature. Check frequently to make sure the incubator is working properly the first day, and continue checking daily thereafter at least four times a day.

(4) About 6 or 7 days after setting, candle the eggs (see Figures 85 and 86) and remove any eggs that are infertile (clear) or have dead germ (cloudy).

(5) At 25 days after setting (Pekin eggs), the eggs are transferred to hatching trays, and if eggs are hatched in a separate machine, moved to the hatcher. Candle and remove eggs with dead embryos. At the time of transfer, the temperature of the hatcher should be set at 37.2°C (99°F) and the humidity set at 65% (31.1°C or 88°F wet bulb). As the hatch progresses, and eggs begin to pip, increase the humidity to 80% (33.9°C or 93°F wet bulb), and increase ventilation openings by about one-half. As the hatch nears completion gradually lower the temperature and humidity so that by the end of the hatch the temperature

Fig. 85. Candled 6 days

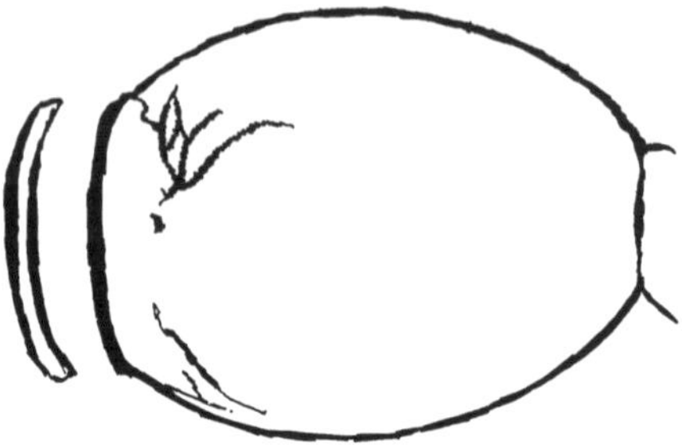

Fig. 86. Fertile egg—6 days

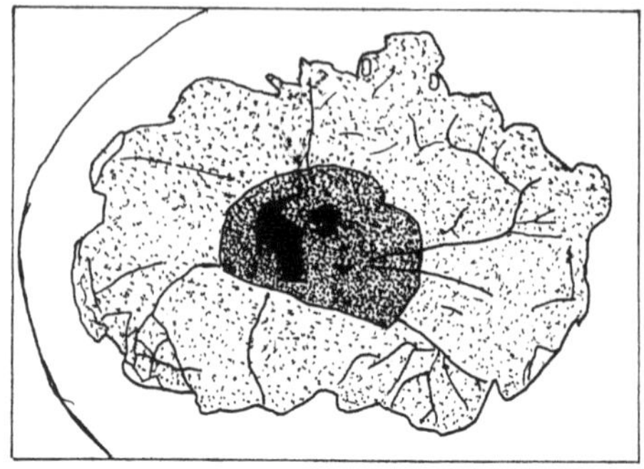

is at 36.1°C (97°F), and the humidity is at 70% (32.2°C or 90°F wet bulb). Vents should be opened to their maximum setting by the end of the hatch. Remove ducklings from the hatcher when 90-95% of them are dry.

Natural incubation. Duck eggs may be hatched naturally by placing them under a broody duck. Muscovy ducks are very good setters, capable of hatching 12-15 duck eggs. The nest box should be located in a clean dry shelter, bedded with suitable litter. Feed and water should be available for the broody duck and for the ducklings when they hatch.

Egg storage. If eggs are stored for a while before they are set, they should be stored at a temperature and humidity level that will minimize deterioration of the egg. For a small number of eggs, storage in a cellar may suffice. Store eggs at about (13°C or 55°F) and 75% relative humidity whenever possible. Store eggs small end down.

Proper water loss during incubation. As the duckling develops inside the egg there is a loss of water from the egg and an increase in the size of the air cell. If the duckling is developing normally, the air cell should occupy about one-third of space inside the egg at 25 days of incubation (Pekin ducks). Weight loss can also be used as a guide. Duck eggs should lose about 14% of their weight at time of setting by 25 days.

E. Duck Nutrition

Poultry nutrition in general is discussed at some length in Section 6 of this book. It is very important for anyone starting out to raise ducks to fully understand the basic fact that no matter how ducks get their food, whether it be by scavenging, eating household food scraps, or consuming home mixed or commercially prepared complete duck rations, the food that is consumed must contain the nutrients (energy, protein, vitamins, minerals, etc.), in an available form (digestible, absorbable, useable), that they need for maintenance, growth and reproduction. As explained in Section 6, feeding practices will depend in part on the number of ducks raised. If only a few ducks are to be kept by a household, and they have access to areas where they can forage, they may be able to survive, grow and lay eggs by consuming table scraps and whatever they can find by foraging. Under such conditions, ducks will likely grow very slowly and produce a small number of eggs. Herded ducks, discussed earlier, will of course perform better, but they require access to large areas, including harvested grain fields, and the care of a herdsman, in order to get enough food.

Feeding a small home flock of ducks without feed concentrates. For small flock owners who have less than 15 ducks and are not able to obtain complete duck feeds, or the ingredients to make complete feeds, the following suggestions are offered. (1) If possible, allow ducks access to areas where the flock can forage for green plants, seeds, insects, insect larvae, earthworms, snails, crabs, mollusks, shrimp, crayfish, small frogs, eels, and other material with some nutritive content. (2) Feed ducks unused garden vegetables, table scraps (bones must be ground), and other food leftovers. (3) If affordable, supplement the flock's diet with any locally available foodstuffs such as bakery waste, rice bran (if not rancid), wheat bran, brewers yeast, distillers grains, broken rice, trash fish (cooked and not putrefied), prawn meal, tapioca chips, palm oil, and similar products.

Feeding larger duck flocks, or for better performance. If keepers of small home flocks want better growth and more eggs they will have to provide supplemental feed. At a minimum they will have to feed some grain. As the size of a home flock is increased above about 10-15 ducks, it becomes more likely that the flock will not be able to get enough food by foraging. It will become necessary to provide the ducks more of their daily diet, and the diet will have to be more complete nutritionally. If more than about 15 ducks are to be kept, or if increased performance is desired, there are a number of choices of feeding practices, such as those listed below.

(1) If complete poultry feeds are available from a commercial feed company, and the feeds are affordable, purchasing feed may be a good choice. Such feeds are usually properly balanced nutritionally and therefore the duck keeper does not have to worry about formulation of the rations, purchasing all the individual ingredients needed and mixing the rations. If duck rations are not available, and chicken rations are, chicken rations will serve as a satisfactory substitute.

(2) If poultry feed concentrates, designed to be fed with grain can be purchased, this too may be a good option. The concentrate contains protein, vitamins and minerals needed to supplement cereal grains, making the total diet complete nutritionally. The grain can sometimes be

purchased from local farmers at a favorable price. A measured amount of concentrate is fed to the flock each day along with grain.

(3) If feed ingredients can be purchased from farmers or feed mills, and are affordable, duck rations can be mixed on the premises, as described in Section 6. This requires some knowledge of feed formulation and purchasing and storing all of the ingredients that go into complete duck rations (see Table 9.6 and Section 6).

Nutrient requirements of ducks. Ducks require the same nutrients as chickens, but in slightly different amounts. Suggested nutrient levels for complete duck rations are listed in Tables 9.4 and 9.5. These levels are set high enough to meet the requirements of all breeds of domestic ducks. Because correct nutrient levels for a particular ration depend on the energy level of that ration (ducks eat progressively more feed as the energy level is lowered and progressively less as it is raised), nutrient requirements are listed in the tables in reference to a particular energy level. For each type of ration, requirements for a high and a low energy ration are given. Requirements for rations with energy levels different from those listed in the tables can be calculated from the ratios of each nutrient to energy, using the values in the tables. Examples of complete duck rations that meet the requirements listed in Tables 9.4 and 9.5, are given in Table 9.6.

Energy. Most of the energy in a typical duck diet comes from the starch in cereal grains. Ducks perform best when their diet contains a high proportion of cereal grains that are high in available energy such as maize, wheat, and sorghum grain. However ducks can perform well on diets that are fairly high in low energy feedstuffs, such as grain by-products, provided the diet is well balanced otherwise. Fortunately, ducks over 4 weeks of age have considerable capacity, when fed a low energy diet, to increase their feed intake and consume enough of the ration to meet their daily nutrient needs. Pekin ducks can grow well on diets containing as little as 2204 kcal/kg (1000 kcal/lb) of metabolizable energy, provided the diet is well balanced. Examples of both high and low energy diets that are properly balanced and support good performance are given in Table 9.6. The example low energy rations utilize low energy feedstuffs like wheat and rice bran, and contain no soybean meal or supplemental amino acids.

Protein. Ducks, like other poultry, do not actually require "protein" but the individual amino acids contained in the proteins in feedstuffs. The proteins in the diet are broken down during digestion to amino acids which are absorbed and used by the duck to make its own body proteins, such as those in muscle and feathers. Certain of these amino acids must be supplied in the diet because the duck cannot make them from other sources. These are called essential amino acids. In formulating feeds for ducks primary attention is paid to meeting the ducks essential amino acid requirements. As noted in Tables 9.4 and 9.5, protein levels that meet the ducks amino acid requirements may vary slightly, depending upon ingredients used in each formulation.

Minerals and vitamins. All of the minerals and vitamins required by ducks, and suggested levels of use in duck rations are listed in Tables 9.4 and 9.5. Close attention should be paid to calcium, phosphorus and sodium when formulating practical rations. Note that the phosphorus requirement is expressed in terms of available phosphorus. A rule of thumb used by nutritionists is that only about 1/3 of the phosphorus in feedstuffs of plant origin (cereal grains, soybean meal, etc.) is available to poultry. Close to 100% of the phosphorus in inorganic (mineral) sources such as dicalcium phosphate is available. Books with information on the amount of available phosphorus in feedstuffs are listed in Section 11. If a duck producer is mixing his own feed, the simplest way, and often the most economical as well, to add vitamins and trace minerals is in the form of commercially prepared premixes. If it is not possible to use prepared premixes, the next best choice is to purchase the vitamin and mineral sources and make your own premixes. If this cannot be done, other suggestions are offered in Section 6.

Fig. 87. Pekin breeders

Feeding developing breeders. As explained earlier, feed intake of developing breeders fed concentrate feeds should be restricted to prevent them from becoming too fat. From the time restriction is begun (usually at about 7 weeks of age), and up until the breeders are mature enough to lay (about 28 weeks of age for Pekin ducks), their daily feed intake should be limited to 60 - 70 % of the amount they would eat if allowed to eat all they want (unrestricted). For example, meat-type Pekin ducks, 7 weeks old, will normally eat about 227 g (0.5 lb.), of a complete pelleted duck ration containing 3086 kcal/kg (1400 kcal/lb.) when fed on an unrestricted basis. This results in an intake of about 700 kcal per day. If we wish to restrict the daily intake of calories to 70% of unrestricted intake, then we should feed 490 kcal (700 x .70 = 490) per breeder per day. This would amount to feeding 159 g (0.35 lb.) per bird per day of a ration containing 3086 kcal/kg. If the developer ration contains a different energy level, the daily feed allotment would have to be adjusted so that the daily intake of 490 kcal remains unchanged (see Table 9.6 for examples). Feed restriction requires hand feeding a weighed amount of feed to the flock each day. Since the ducks are very hungry at the time of feeding, the feed must be spread out so that all ducks have a chance to eat. Feed can be spread out in long wooden troughs, on a cement slab or on the ground if the area is dry and clean. A few ducks from the flock should be handled and weighed occasionally to determine if they are gaining too much or too little weight. If breeders are developing normally they should have good body conformation and bone development but should not have excess fat, which is very easy to feel in the abdomen. If developers are getting too fat, their daily feed allotment should be reduced. If they are too light, the daily feed should be increased.

Feeding laying breeders. Nutrient levels that duck breeder layer rations should contain are listed in Table 9.5, and examples of breeder rations in Table 9.6. Layer rations contain a higher level of calcium than other duck rations to meet the laying duck's need for additional calcium for egg shell formation. A level of 3.00% of the diet is adequate for most breeds of ducks including high egg producing breeds. When enough calcium is included in the ration, it is not necessary to feed oyster shells in addition. However, it will do no harm to make oyster shells available, as is the practice on some duck farms.

Feedstuffs. Some guidelines to the choice of feedstuffs are given in Section 6, which apply to ducks as well as chickens. Some feed ingredients contain substances that are toxic to ducks, and should not be included in duck rations. Groundnut meal (peanut meal) is often contaminated with aflatoxin, a toxin to which ducks are highly sensitive. Groundnut meal should not be used unless tests have proven it to be free of aflatoxin. Rapeseed meal is another feedstuff that contains toxins (goitrogens and isothiocyanates) which are very harmful to ducks. There are some improved varieties of rapeseed meals that are much lower in these toxins. Meal from these improved varieties is called canola meal in Canada. However even canola meal should be tested with ducks before it is used in duck feeds.

Feed quality. Duck keepers, like all animal producers, must be diligent in making sure that all the feedstuffs fed to their ducks are of good quality. One of the most common causes of poor quality is failure to dry grains and other feedstuffs properly before storage. If grains that are too high in moisture are stored, without turning or aeration, the grain will heat up and mold and some of its nutritive value will be destroyed. As explained later, some molds may produce toxins that are particularly harmful to ducks. Make sure that the grains and other feedstuffs used in duck feeds were properly dried and are free of molds and other contamination. If table scraps, bakery waste, wet mash or other feeds high in moisture are fed, feed only what ducks will clean up in a day. If such feeds remain in troughs longer, they will likely become moldy. Feedstuffs that are to be stored for very long should contain no more than 10-12% moisture. Do not feed ducks feedstuffs containing rancid fat. Rice bran, if not properly cured and stored, has a tendency to become rancid. Avoid feeding fish and fish products that are putrefied.

Water. Plenty of clean drinking water should be available to ducks at least 8-12 hours per day. In some management systems it is advantageous to shut off feed and water at night to help maintain litter inside buildings in a dry condition. This applies to breeder ducks or market ducks over 3 weeks of age. If done properly, this practice is not harmful and has no effect on performance. Ducks do not require water for swimming in order to grow and reproduce normally. However, providing some water for wading or swimming can be beneficial, especially in hot climates. Ducks can expel excess heat through their bill and feet when allowed contact with water that is appreciably lower in temperature than their body temperature 41.7°C (107°F). Water temperatures of 10-21°C (50-70°F) are ideal for ducks.

Mash or Pellets. It is a well established fact that ducks grow faster when fed pelleted rations than when the same ration is fed in mash form. Pelleting all duck rations is common practice in commercial duck production. However, pelleted feeds are not available in many areas of the world, and it may be difficult for small flock owners to get their home mixed feed pelleted. Pelleting equipment is quite expensive and can usually be afforded only by commercial feed mills and large commercial farms. The problem with feeding dry mash to ducks is that it forms a sticky paste when mixed with saliva, cakes and accumulates on the outer ridges of the mouth. In attempting to free their bills of caked feed, ducks make frequent trips to water to wash their bills, causing feed wastage. Feeding mash also reduces feed intake, and in the case of market ducks, reduces their growth rate. For small flock owners who are not able to pellet their duck feeds, one solution to the problems of feeding dry mash is to feed wet mash. Water is mixed with the mash just before feeding. Enough water is added to form a thick mush without making it watery. Mix only what ducks will clean up within a few hours or at most a day. The mixing methods described in Section 6 can be applied to mixing wet mash.

Pellet size. When pelleted feeds are fed to ducks it is important to avoid feeding pellets that are too large in diameter or too long for ducklings to swallow. For newly hatched ducklings, pellets should be no larger in diameter than 4.0 mm (5/32 inch), and no longer than 7.94 mm (5/16 inch). After about two weeks of age, Pekin ducklings can consume pellets 4.76 mm (3/16 inch) in diameter, and approximately 12.7 mm (1/2 inch) in length, without difficulty.

F. Duck Processing

Ducks are killed and dressed in a manner similar to chickens. This can be done on a small scale with simple equipment or in a modern poultry dressing plant where much of the work is mechanized. Ducks are killed by placing them head down in funnels, or hanging them by the legs in shackles, and cutting the jugular vein and carotid artery on one side of the neck next to the jaw. Ducks are scalded in water at (57-63°C; 135-145°F) for 2-3 minutes. Ducks may be plucked by hand or with the assistance of a picking machine (a rotating drum with rubber "fingers"). To aid in the removal of pin feathers, ducks, unlike chickens, are dipped in melted wax, followed by submersion in cold water to harden the wax. The wax is then stripped off, bringing with it hard to remove pinfeathers (Figure 88). Since duck down feathers can be used in making pillows, comforters and down-filled jackets, raw feathers are sometimes washed, dried and baled for sale to feather processors.

Fig. 88. Stripping wax

G. Diseases of Ducks

Disease prevention, discussed in Section 7, is just as important in raising ducks as in raising other poultry. The major diseases of domestic ducks are described briefly below. For more information, see the sources listed Section 11.

Duck Virus Hepatitis. Duck virus hepatitis is a highly contagious, highly fatal disease of young ducklings, 1-28 days of age. Ducklings are most susceptible at the younger ages and gradually become more resistant as they grow older. The disease is rarely seen in ducklings over 4 weeks of age. The onset of the disease is very rapid, it spreads quickly through the flock and may cause up to 90% mortality. Sick ducklings develop spasmodic contractions of their legs and die within an hour in a typical "arched-backward" position. The liver is enlarged and shows hemorrhagic (blood from ruptured blood vessels) spots. Preventing introduction of this disease, keeping age groups isolated, and vaccination of breeder ducks with an attenuated live virus duck hepatitis vaccine (to produce parentally immune ducklings) are effective control measures.

Duck Viral Enteritis (Duck Plague). Duck Viral Enteritis is an acute, contagious, highly fatal disease of waterfowl caused by a herpes virus.

In domestic ducks this disease is most likely to affect mature ducks, but is also seen in young ducks. Affected birds show sluggishness, ruffled feathers, greenish-yellow diarrhea that is sometime blood-stained. Dead birds often have blood-stained feathers around the vent and blood dripping from the nostrils. Hemorrhages may be found in tissues throughout the body. Eruptive lesions of the mucous lining of the esophagus and intestine are characteristic signs of the disease. Regular immunization of breeder ducks with an attenuated live duck viral enteritis vaccine provides adequate protection.

Riemerella Anatipestifer Infection. This bacterial disease of ducks has also been called Pasteurella Anatipestifer Infection, Infectious Serositis and New Duck Disease. Anatipestifer infection causes high mortality, weight loss and diseased carcasses at slaughter time in market ducks. In the acute form, listlessness, eye discharge and diarrhea are commonly seen. Ducks show incoordination, shaking of the head and twisted neck. Birds are commonly found on their backs, paddling their legs. Typical lesions found in dead birds are infected air sacs and membranes covering the heart, liver, brain and spinal cord. The liver is enlarged and covered with a fibrinous membrane. Preventive management and inoculation with bacterins are effective means of control.

Cholera. Avian cholera, caused by the bacterium Pasteurella multocida is an important disease of domestic ducks, and is an especially troublesome disease of ducks in some parts of Asia. This disease is associated with poor sanitation, and standing water in duck pens. Symptoms include loss of appetite, mucous discharge from the mouth, diarrhea, and in breeder ducks, labored breathing. Lesions found in dead birds include hemorrhages on heart muscle, mesentery and abdominal fat. The liver is enlarged, copper colored and friable (easily crumbled). Pinpoint whitish spots may be seen on the liver. Good sanitation practices go a long way toward preventing this disease. Effective drugs (given in the feed) include Sulfaquinoxaline (0.05%), Sulfadimethoxine-ormetoprim (0.02-0.04%) and Chlortetracycline (0.044%).

Colibacillosis. This common infection of poultry caused by *Escherichia coli,* causes reduced hatchability, infection of the yolk sac (omphalitis),and a septicemia (bacterial invasion of bloodstream) in ducks 2-8 weeks of age and salpingitis and peritonitis in breeder ducks. In market ducks, *E. coli* infection produces lesions very similar to those seen in Anatipestifer infection (see above). Good sanitation and management are important preventive measures. Sulfadimethoxine-ormetoprim (0.02-0.08% in feed) is helpful in controlling this disease.

Aspergillosis. This condition occurs when ducks inhale spores produced by the mold Aspergillus (*Aspergillus fumigatus* is the common species) that grows on straw or damp feed. These inhaled spores cause multiple nodules or plaques in the lungs and airsacs. Common signs include gasping, listlessness and dehydration. This disease is not to be confused with aflatoxin poisoning described below. The best solution to aspergillosis is to avoid using moldy straw and preventing feed from getting wet and moldy.

H. Toxins

Aflatoxin poisoning. Molds that grow on cereal grains and oilseeds before and after harvest produce a number of toxins that are particularly harmful to ducks. By far the most toxic of these substance is a group of toxins called aflatoxins. Aflatoxins are produced by the molds *Aspergillus flavus* and *Aspergillus parasiticus.* Ducks are highly susceptible to these toxins. Very small amounts will cause high mortality. Wet harvest conditions encourage the growth of this mold.

Botulism. Ducks that have access to stagnant ponds or other areas where decaying organic matter (animal carcasses, in particular) is found may consume toxins produced by the microorganism *Clostridium botulinum.* This happens when temperature and other conditions are right for the growth of this anaerobic (lives without oxygen) spore-forming microorganism. Botulism causes a progressive flaccid (limp) paralysis of the neck (limberneck), legs and wings. Affected ducks usually die in a coma within 24-48 hours.

Castor bean poisoning. Incidents of high death losses in wild ducks, due to consuming castor beans (*Ricinus communis*) have been reported in Texas. Castor beans contain ricin, a toxalbumin known to cause toxicity in humans and domestic animals.

Insecticides, rodenticides. Duck keepers should take care not to use insect sprays or rat and mouse poisons, that are known to be harmful to ducks, in areas where ducks are kept. Some insect sprays are highly toxic to ducks, such as parathion and diazinon. Always read the directions on the insecticide container carefully before using around ducks. Rat poisons that contain Warfarin, an anticoagulant, can cause ducks to bleed to death, if they consume the poison.

TABLE 9.4 Suggested Nutrient Levels for Duck Rations

Nutrient	*Starter, 0-14 days*		*Grower-finisher, 14-49 day*	
(% or amt/kg diet)	*High Energy*	*Low Energy*	*High Energy*	*Low Energy*
Met energy, kcal/kg[1]	3086	2646	3086	2646
Met energy, kcal/lb[1]	1400	1200	1400	1200
Linoleic acid, %	1.00	0.87	1.00	0.88
Protein, % (approx.)[2]	22.0	19.1	16.1	14.0

TABLE 9.4 (continued)

Nutrient	*Starter, 0-14 days*		*Grower-finisher, 14-49 day*	
(% or amt/kg diet)	*High Energy*	*Low Energy*	*High Energy*	*Low Energy*
Lysine, %	1.20	1.04	0.80	0.70
Methionine, %	0.47	0.41	0.35	0.31
Meth + cystine, %	0.80	0.70	0.60	0.52
Tryptophan, %	0.23	0.20	0.20	0.17
Arginine, %	1.20	1.04	1.00	0.88
Valine, %	0.88	0.77	0.80	0.70
Threonine, %	0.80	0.70	0.60	0.52
Histidine, %	0.44	0.38	0.35	0.31
Isoleucine, %	0.88	0.77	0.70	0.61
Leucine, %	1.40	1.22	1.30	1.14
Phenylalanine, %	0.80	0.70	0.70	0.61
Phenyla+tyrosine,%	1.50	1.31	1.30	1.14
Calcium, %	0.70	0.57	0.65	0.57
Phosphorus, avail,%	0.40	0.35	0.35	0.31
Sodium, %	0.15	0.13	0.14	0.12
Chlorine, %	0.16	0.14	0.14	0.12
Potassium, %	0.60	0.52	0.60	0.52
Magnesium, mg/kg	500	435	500	435
Manganese, mg/kg	50	44	40	35
Zinc, mg/kg	60	52	60	52
Selenium, mg/kg	0.15	0.13	0.15	0.13
Iodine, mg/kg	0.40	0.35	0.40	0.35
Iron, mg/kg	80	70	80	70
Copper, mg/kg	8	7	6	5
Vitamin A, IU/kg	5000	4350	4000	3480
Vitamin D3, IU/kg	600	522	500	435
Vitamin E, IU/kg	25	22	20	17
Vitamin K, IU/kg	2	2	1	1
Choline, mg/kg	1300	1130	1000	870
Riboflavin, mg/kg	4	4	3	3
Niacin, mg/kg	50	44	40	35
Pantothenic a.,mg/kg	12	10	10	9
Thiamin, mg/kg	2	2	2	2
Vitamin B_{12}, mg/kg	.01	0.01	0.005	0.004
Pyridoxine, mg/kg	3	3	3	3
Folic acid, mg/kg	0.50	0.50	0.25	0.22
Biotin, mg/kg	0.15	0.13	0.10	0.09

[1] Metabolizable energy level on which all nutrient requirements in column are based.
[2] Formulate ration to meet amino acid levels, protein level may vary slightly.

TABLE 9.5 Suggested Nutrient Levels for Duck Rations

Nutrient	*Breeder-developer*		*Breeder-layer*	
(% or amt/kg diet)	*High Energy*	*Low Energy*	*High Energy*	*Low Energy*
Met energy, kcal/kg[1]	2866	2205	2866	2646
Met energy, kcal/lb[1]	1300	1000	1300	1200
Linoleic acid, %	1.10	0.80	1.00	0.93
Protein, % (approx.)[2]	17.6	12.8	17.5	16.2
Lysine, %	0.88	0.64	0.80	0.74
Methionine, %	0.38	0.28	0.40	0.37
Meth + cystine, %	0.66	0.48	0.65	0.60
Tryptophan, %	0.22	0.16	0.20	0.18
Arginine, %	1.10	0.80	0.87	0.80
Valine, %	0.88	0.64	0.75	0.69
Threonine, %	0.66	0.48	0.58	0.54
Histidine, %	0.38	0.28	0.37	0.34
Isoleucine, %	0.77	0.56	0.72	0.67
Leucine, %	1.43	1.04	1.22	1.13
Phenylalanine, %	0.77	0.56	0.70	0.65
Phenyla+tyrosine,%	1.43	1.04	1.10	1.02
Calcium, %	0.75	0.55	3.00	2.77
Phosphorus, avail,%	0.40	0.29	0.40	0.37
Sodium, %	0.16	0.12	0.15	0.14
Chlorine, %	0.15	0.11	0.14	0.13
Potassium, %	0.66	0.48	0.60	0.56
Magnesium, mg/kg	550	400	500	465
Manganese, mg/kg	44	32	40	37
Zinc, mg/kg	66	48	60	56
Selenium, mg/kg	0.16	0.12	0.15	0.14
Iodine, mg/kg	0.44	0.32	0.40	0.37
Iron, mg/kg	88	64	70	65
Copper, mg/kg	7	5	6	5.6
Vitamin A, IU/kg	4400	3200	6000	5580
Vitamin D_3, IU/kg	550	400	600	558
Vitamin E, IU/kg	22	16	30	28
Vitamin K, IU/kg	2	2	2	2
Choline, mg/kg	1100	800	1000	930
Riboflavin, mg/kg	4	3	4	4
Niacin, mg/kg	44	32	50	46
Pantothenic a.,mg/kg	11	8	12	11
Thiamin, mg/kg	2	2	2	2
Vitamin B12, mg/kg	0.01	0.01	0.01	0.01
Pyridoxine, mg/kg	3	2	3	3

TABLE 9.5 (continued)

Nutrient	Breeder-developer		Breeder-layer	
(% or amt/kg diet)	High Energy	Low Energy	High Energy	Low Energy
Folic acid, mg/kg	0.30	0.22	0.50	0.50
Biotin, mg/kg	0.11	0.08	0.15	0.14

[1] Metabolizable energy level on which all nutrient requirements in column are based.

[2] Formulate ration to meet amino acid levels, protein level may vary slightly.

TABLE 9.6 Examples of Complete Duck Rations

	Starter, 0-14 days		Grow-finisher, 14-49 day	
Ingredient	Hi-Energy 3086kcal/kg	Low-Energy 2646kcal/kg	Hi-Energy 3086kcal/kg	Low-Energy 2646kcal/kg
	%	%	%	%
Ground Maize	64.44	35.37	73.37	47.81
Sorghum grain (milo)	—	10.00	—	—
Wheat middlings	—	4.39	3.04	15.63
Wheat bran	—	20.00	—	15.06
Rice bran	—	10.00	—	10.00
Soybean meal, 48.5%	26.54	—	17.20	—
Maize gluten meal	—	1.63	—	—
Sunflower meal	—	—	—	5.00
Fish meal, menhaden	1.50	8.50	—	5.00
Meat & bone meal	5.00	—	4.87	—
Brewers yeast	—	9.17	—	—
Fat, animl-veg blend	1.51	—	0.56	—
Limestone	0.07	0.21	0.11	0.57
Dicalcium phosphate	—	—	—	0.18
DL-methionine	0.11	—	0.05	—
Salt	0.22	0.15	0.21	0.14

TABLE 9.6 (continued)

	Starter, 0-14 days		*Grow-finisher, 14-49 day*	
Ingredient	***Hi-Energy 3086kcal/kg***	***Low-Energy 2646kcal/kg***	***Hi-Energy 3086kcal/kg***	***Low-Energy 2646kcal/kg***
	%	%	%	%
Trace mineral mix	0.10	0.10	0.10	0.10
Vitamin mix	0.50	0.50	0.50	0.50
TOTAL	100.00	100.00	100.00	100.00

	Breeder-developer[1]		*Breeder-layer*	
Ingredient	***Hi-Energy 2866kcal/kg***	***Low-Energy 2205kcal/kg***	***Hi-Energy 2866kcal/kg***	***Low-Energy 2646kcal/kg***
	%	%	%	%
Ground Maize	61.18	17.88	62.49	53.92
Sorghum grain (milo)	—	15.00	—	—
Wheat middlings	14.33	—	5.62	6.14
Wheat bran	—	50.51	—	10.00
Rice bran	—	10.00	—	5.00
Soybean meal, 48.5%	19.50	—	16.68	—
Sunflower meal	—	0.47	—	3.18
Fish meal (menhaden)	—	4.75	1.00	8.00
Meat & bone meal	2.00	—	5.00	2.00
Brewers yeast	—	—	—	5.00
Fat, animl-veg blend	0.38	—	1.50	—
Limestone	0.72	0.62	6.71	6.01
Dicalcium phosphate	0.96	—	0.09	—
DL-methionine	0.08	—	0.09	—
Salt	0.26	0.16	0.22	0.16
Trace mineral mix	0.10	0.10	0.10	0.10
Vitamin mix	0.50	0.50	0.50	0.50
TOTAL	100.00	100.00	100.00	100.00
Feed/duck/ day, g[1]	171(490kcal)	222(490kcal)	—	—

[1] To be fed on a restricted basis, at 60-70% of unrestricted daily consumption, see text. Daily allotments shown are for Pekin ducks, 7-28 weeks of age.

Section **10**

Other Kinds of Poultry

A. Geese. Domestic geese are raised for their meat, large eggs, feathers, and sometimes their ability to weed crops or act as watch dogs. They are large, hardy, aggressive, noisy birds that need little attention when mature. Geese may live for 30 years. They are preferred by some because they may be hardier than ducks and subsist largely on grass. Of nine recognized breeds, African and Chinese do best in the tropics. No brooder is needed for goslings, but they should be kept warm and dry for the first few weeks. In the presence of the mother no outside shelter is required unless weather is extremely cold. Lock up goslings at night to protect them from predators. Although they like to swim, they do not need water to breed. Ideally, starter feed for geese should contain 20 to 22 percent protein in pellet form. They can go to range at 2 to 6 weeks. If pasture is plentiful, geese do not need more than I kg (2 lbs.) of supplemental feed per bird in a week for satisfactory growth.

B. Guinea Fowl. Meat of guinea fowl is prized in some countries for its wild gamebird-like taste. They are valued for watch dog services on some farms. Guineas are wilder than other poultry, hide their nests, and do not take to confinement rearing as well as chickens. They should not be disturbed while laying and 2-3 eggs should be left in the nest to attract the females back again. A broody hen makes a better mother than a female guinea fowl. Keets (hatchlings) are hatched after 28 days of incubation. Ideally they are started on 24% protein diets which can be reduced to 15% protein in the grower mash. Eggs are somewhat smaller than hen's eggs and their production is usually limited to less than 100 per year. In some areas a market demand for guinea eggs has developed. If fed late in the afternoon they may return to shelter for the night. At 18 weeks of age they may weigh up to 1.5 kg (3-1/2 lbs.) when marketed for meat.

C. Turkeys. Originally a hardy bird of the North American forests, domesticated turkeys have been bred for large size and as an inexpensive meat source. The two most common breeds, Bronze and Whites, have difficulty in reproducing because of the heavy weight of the males and their poor fertility. Poults are prone to panic and can break their necks by running into fences, or they may pile up and suffocate each other. They may drown themselves in a low dish of water or jump into a bucket of water. They may starve if not encouraged to start eating. A caretaker can attract them to food by finger tapping on a board making a pecking sound near the feed. Bright colors such as red will attract them to feeders or waterers.

Turkeys frequently suffer from a disease known as blackhead. Signs include yellow feces with large rounded lesions on the liver. The causative agent, a protozoan parasite, also mildly affects chickens. The protozoan parasite is usually transmitted via a cecal worm—a nematode about one centimeter long which lives in the cecum, or blind sac at the lower end of the intestine. Heavy losses in turkeys are prevented by following the rule: "never raise turkeys and chickens together". Coccidiosis, although due to different species from chickens, may require preventive methods similar to those of chickens.

One gobbler (tom) is needed for each four or five hens. However, with artificial insemination, only one is needed for each ten hens. Poults need 28% protein in the starter feed which may be reduced to 20% in grower and 15% in finisher feed.

Until recently turkeys have been known as a luxury bird raised largely for festive occasions. Their large size, 8 to 15 kg (18-35 lbs.) has limited carcass storage without refrigeration. Development of smaller breeds, confinement rearing, together with deboning and cut up processing, has greatly increased market demand. They are now competitive with beef and broilers to satisfy a growing demand for meat.

D. Pigeons. Once established, a pigeon colony can be practically self-sustaining. Larger breeds such as White King and Giant Homer are preferred to the common Rock Pigeon which throngs city parks.

A breeding pair will produce ten or more squabs (young pigeons) each year, two at a time. Squabs are usually eaten when 25 to 30 days old when completely feathered, but before they begin to fly. Any waterproof house will do for housing if protected against cats, dogs, snakes and rats. In Egypt, especially designed shelters, or cotes, are built for recovery of the manure, which is more valuable for fertilizer than the harvest of squabs.

E. Quails. Quail meat and quail eggs are sometimes featured on menus of fancy hotels to satisfy a demand for a wild-bird taste. Both the bob white quail and the much smaller Japanese quail (Coturnix coturnix japonica) are raised commercially. They are generally reared in confined batteries. The Japanese quail are more prolific since they grow to breeding maturity in 6 weeks.

Section 11

Other Information

A. Glossary—Definitions of Poultry Related Terminology

Abscess—a localized collection of pus within a cavity usually surrounded by inflamed tissue.

Acute—a short and severe course as applied to a disease.

Additive—an ingredient added to a basic feed mix.

Agglutination test—serological test to determine antibody levels; example: used to detect *Salmonella pullorum,* the agent causing pullorum disease.

Air cell—air space in egg between shell membranes due to CO_2 and H_2O evaporation.

Air sacs—part of bird respiratory system (lungs, air sacs, bones).

Albumen—egg white.

Albumin—class of water soluble proteins

Allantois—extra-embryonic membrane in an egg.

Amino Acid—one of the organic acids which form the building blocks from which proteins are built.

Anemia—a condition in which there is a deficiency in hemoglobin content or number of red blood corpuscles.

Antibiotic—a chemical substance produced by microorganisms which has the ability to destroy bacteria or other microorganisms.

Antigen—a substance that gives rise to an antibody when introduced into blood or tissue.

Antioxidant—A chemical feed additive used to protect certain nutrients from oxidation.

Antitoxin—a substance present in the blood serum of an animal that has become immune to a certain toxin.

"As hatched" = "Straight run"—unsexed birds from a hatchery; half males and half females, generally used with broilers—rarely layers.

Aspergillosis—a diseased condition cause by a specific fungus present in damp litter.

Attenuate—to weaken.

Bacteria—one-celled, microscopic plants some of which are beneficial while others cause diseases.

Bacterin—similar to a vaccine, but protects against bacterial diseases; produced from dead organisms.

Balut or Embryo Egg—made from a developing chicken or duck embryo.

Beak trimming—removal of a portion of the beak by use of a hot iron or scissors. Also called 'debeaking'.

Billing—a pecking action of poultry spilling out feed from a feeder.

Breed—a selected, closely related group of birds having a common ancestry.

Broiler—a fryer chicken marketed for meat at 6-12 weeks of age.

Brooder—heat distributing equipment used for young chicks to keep from getting chilled.

Brooder Ring (hover guard or chick guard)—a small barrier used around a brooder to keep chicks close to heat to prevent chilling and piling up.

Broody—the condition of a hen which is sitting on a clutch of eggs.

Bursa = bursa of Fabricious—A sac or pouch in the connective tissue located just inside the vent.

Candle—to examine the contents of an egg by beaming a strong light source (from a candler) through the contents.

Cannibalism—a bad habit of chickens pecking on themselves or other birds. Feather pulling is often the first sign of trouble.

Cecum—a blind pouch at the junction of the small and large intestine.

Chick—young, unsexed chicken.

Chigger—larval mites which cause intense itching in humans, blisters on turkeys.

Chronic—a disease condition of long duration in contrast to acute.

Cloaca—the common chamber into which the intestinal, generative, and urinary tracts discharge.

Clutch—a group of eggs laid in a series without missing a day; a group of chicks hatched by a broody hen.

Coccidiosis—a protozoan disease causing losses in chickens or turkeys. The parasites are called coccidia.

Coccidiostat (anticoccidial)—A drug which inhibits development of coccidia usually administered in the feed.

Cock—male, 1 year or older.

Cockerel—male, less than 1 year of age.

Comb—a fleshy head projection.

Concentrate—A feedstuff especially rich in distinctive nutrients.

Confinement—A system of management wherein birds are kept in an enclosure.

Congestion—an excessive accumulation of blood.

Crest—tuft of feathers on head.

Crop—an enlargement of the esophagus in front of the chest.

Crumbles—a preparation consisting of medium size granules prepared from pelleted feed.

Cull—a bird no longer useful for production purposes. Also used as a verb: to cull.

Custom hatching—hatching for hire—once popular—now discontinued in the USA, to prevent disease transmission in the hatchery.

Debeak—removal of a portion of the beak by use of a hot iron or scissors. Also called 'beak trimming'.

Disease—any deviation from a condition of health.

Disinfectant—a chemical agent used in killing microorganisms.

Dub or dubbing—comb removal—to prevent physical (cage) or damage from freezing.

Ducks—females; Drakes = males; Ducklings = baby ducks

Dusting—natural activity of birds rolling in dirt or dust to remove parasites.

Extension—adult education, especially farming methods.

Feces—droppings excreted from the digestive tract.

Feed efficiency—the ratio derived by dividing the weight of feed consumed by the body weight of the bird.

Floor eggs—eggs laid on floor versus cage or nest—undesirable.

Floor pen—a pen where chicks are usually reared on litter in contrast to cage operations.

Floor layer—hen which lays floor eggs.

Fluff—Soft downy feathers in masses on certain parts of fowls.

Fryer—Broiler—young chicken, 6-12 weeks of age, of either sex that has tender meat, a soft pliable smooth skin and a flexible breastbone.

Full-feeding (ad libitum)—having feed in front of animals at all times.

Fumigate—a procedure using fumes to disinfect a given area.

Fungi = Molds—lower forms of plant life which invade tissue such as combs of poultry. They may produce toxins in feedstuffs.

Germinal disc—little white spot on yolk = site the new embryo.

Giblets—edible viscera (gizzard, liver, heart and neck).

Hackles—narrow pointed feathers on neck.

Haugh units—method of measuring internal egg quality on broken-out eggs (wt. of egg and albumen height).

Hock Joint—Joint in lower leg just above the foot.

Hover or brooder—canopy or cover of a brooding device that supplies or retains heat for brooding chicks.

Hen—female greater than 1 year of age (or in lay).

Hybrid—offspring from parents of different breeds, varieties or species.

Hybrid Vigor, (heterosis) an increased vigor or capacity for growth often occurring in crossbred animals or plants.

Immunity—the state of being protected against a disease.

Incubation—the process of maintaining eggs under conditions favorable for the development of the embryos.

Infection—a contamination with any disease-producing substance such as bacteria or viruses.

Inflammation—a condition in which tissues manifest swelling, redness, heat and pain.

Integration—combinations of different poultry operations under one management (e.g., hatching, rearing, processing).

Internal layer—an egg released from the ovary which misses the journey down the oviduct; yolk is retained in the body cavity.

Keel bone—the breast bone to which breast muscles are attached.

Lesion—an abnormal structural change in an organ due to injury or disease.

Litter—absorbent material placed on the floor of a chicken house.

Marek's disease—a cancer-type virus disease affecting younger chicks.

Metabolic—the chemical and physical process going on in living tissues.

Meat spots—eggs which contain pieces of hen tissue in the albumen.

Nares—nostrils.

Necropsy or autopsy—to perform a post mortem examination.

Newcastle disease—a serious virus disease causing high mortality.

Nick—genetic combinations of male and female lines which produce desired characteristics in offspring.

Oil gland = preen gland = uropygial gland—at base of tail used by bird to oil feathers, preening; removed during processing.

Oocyst—an egg-like structure which is the infectious stage causing coccidiosis.

Parasites—live organisms which invade tissues or live at the expense of other organisms sometimes causing diseases.

Pathogen—a disease producing organism.

Peck order—established social order in flocks—insures peace.

Pellet—a formed large granule of mixed feed prepared by heat treatment in the feed mill.

Pipping—embryo breaks shell and starts to hatch, egg is then called a pip.

Post mortem = autopsy = necropsy—examination of a dead bird in an attempt to determine the cause of death.

Poult—unsexed turkey chick.

Premix—a blended mixture of such ingredients as vitamins or minerals combined with a small quantities of a carrier to facilitate transfer to a bulk mixer.

Protozoa—one-celled organisms belonging to the animal kingdom.

Proventriculus—the glandular stomach of birds.

Pullet—young hen less than one year old.

Ration—the diet of an animal.

Replacement stock—Young birds brought in to replace worn-out breeders or layers.

Rickets—a disease caused by vitamin D deficiency resulting in soft bone structure.

Roaster—a young chicken suitable for roasting.

Rooster—the male of the domestic fowl, a cock.

Salmonella—a genus of bacteria containing species which cause digestive disturbances in humans or in birds (eg. pullorum disease).

Sanitation—the adoption of practices to make or keep clean; to eliminate unhealthy elements.

Self-feeder—an automatic feed hopper that lets feed down as the bird eats.

Set—to start eggs for incubation under a hen or in an incubator.

Sexer—an expert at determining the sex usually done by examining the vent on day-old birds using a strong light.

Sexing—determining the sex, usually done on day-old birds.

Sign—Any objective evidence of an abnormality such as coughing, sneezing or vomiting, a symptom.

Species—a distinctive category in animal or plant classification constituting a subclass under genus.

Spent hens—hens that have completed their useful life.

Straight run—Unsexed chicks or poults as they come from the hatchery.

Strain—A closely related group of birds within the same breed.

Stress—a stimulus which may lower resistance to disease such as chilling, moving, or loud or frightening sounds.

Supplement—a feed additive used to improve the nutritive balance.

Toxin—a poisonous substance produced by certain bacteria or fungi, and proteins from animal or plant sources.

Trachea—the thin-walled air tube from the mouth to lungs.

Tumor—new tissue produced by abnormal cellular growth which is unrelated physiologically to neighboring cells.

Vaccine—usually a suspension of killed or modified live organisms used to protect against a specific infectious disease.

Vector—a carrier of a disease-producing microorganism, as an insect.

Vent—the common outlet for the urinary and digestive tracts of birds.

Virus—an infectious agent too small to detect with light microscopes, but usually demonstrated under electron microscopes; many diseases of poultry are caused by viruses.

Viscera—the internal organs in the body cavity.

Vitamin—an organic compound required in the diet that plays an essential role in body metabolism.

Wattles—a fleshy often bright colored piece of skin hanging from the throat of birds.

B. Useful References.

GENERAL

a. Anonymous, Undated. Arbor Acres Broiler Breeders Feeding and Management. Arbor Acres Farm, Inc., Glastonbury, Connecticut 06033, USA.

b. Ash, W. A., 1976. Raising Ducks. Farmers' Bulletin No. 2215. U.S.D.A., Superintendent of Documents, U.S. Govt. Printing Office, Washington, D.C.

c. Bishop, John P., DVM, PhD., Chick Production and Health, Guide to Improving Small-farm Flocks in the Tropics. Available from the Heifer Project International, P.O. Box 8058, Little Rock, AR 72203 U.S.A., or info@heifer.org, or telephone 1-800-422-0474.

d. Ensiminger, M. 1980. Poultry Science. 2nd Ed. Interstate Printers & Publishers, Danville, Illinois 61832 U.S.A.

e. Hetzel, D.J.S., 1986. Domestic ducks: An historical perspective. In Duck Production Science and World Practice, pp. 1-5.

f. Holderread, D., 1987. Raising the Home Duck Flock. A Garden Way Publishing Book. Storey Communications Inc., Pownal, Vermont 05261.

g. Nesheim, M. C., Austic, R., Card, L. 1979. Poultry Production. 12th ed., Lea & Febiger, Box 3024, 200 Chesterfield Parkway, Malvern, PA 19355-9725. Used as the beginning textbook in many poultry departments. Presents orientation on commercial poultry operations.

h. North, Mack O. 1986. Commercial Chicken Production Manual. 3rd ed. Contains extensive data on development of commercial poultry enterprises. 115 Fifth Avenue, New York, NY 10003, U.S.A.

i. Orr, H.L., 1969. Duck and Goose Raising. Publication 532. Ontario Department of Agric. and Food, Parliament Building, Toronto, Canada.

j. Scott, M.L., and Dean, W.F., 1991. Nutrition and Management of Ducks. M.L. Scott, Ithaca, NY 14853.

Elementary, or Nontechnical

a. French, Kenneth M. 1981. Practical Poultry Raising. Peace Corps Information and Exchange. Manual Number 11. 1990 R St. N.W., Washington, DC 20520. For sale by National Public Information Service, Springfield, VA 22161. Excellent material on how to start a small scale poultry enterprise. Describes the role of extension workers introducing people to poultry development.

b. Florea, J. H. 1977. ABC of Poultry Raising: A Complete Guide for the Beginner or Expert. 2nd Ed. Dover Publications Inc., NY. 142 p.

c. Hy-Line™. Red Book: Management and disease control. For sale Hy-Line Indian River Co., Box 65190, West Des Moines, IA 50265.

Nutrition and Feed Formulation

a. Tables of Feed Compositions, Publication 1684, National Academy of Sciences, National Research Council, Washington, DC. 1964. These tables show average nutritive contents of most available poultry feedstuffs found in different parts of the world.

b. Nutrient Requirements of Poultry. 8th Edition, 1984. National Research Council, Washington, DC 20418. Deficiency diseases are herein described.

c. Scott, M. L., Nesheim, M. C., and Young, R. J. Nutrition of the Chicken. 3rd Ed. 1982. published by M. L. Scott and Associates, Ithaca, NY. A textbook describing feed formulation to meet the nutritional requirements.

d. Halbrook, E. R., W. J. Mueller, W. W. Thomann, H. Engler, 1965. FAO Agricultural Development Paper No. 82. Poultry Feeding in Tropical and Subtropical Countries. Food and Agriculture Organization of the United Nations, Rome.

Diseases

a. i) Copeland, J. W. Newcastle disease in poultry. A new food pellet vaccine. Australian Centre for International Agricultural Research. Canberra, A. C. T. 2501. A convenient method of administering vaccine to backyard flocks.

b. Hofstad, M. S., Calnek, B. W., Helmboldt, C. F., Reid, W. M., Yoder, H. W., Jr. 1984. Diseases of Poultry, 8th ed. Iowa State University Press, Ames, IA 50010. 948 pp. Excellent reference work for advanced students.

c. Salsbury Manual of Poultry Diseases. 7th ed. Solvay Animal Health, Inc., Mendota Heights, MN 55120. Excellent colored illustrations and description of 52 diseases in English, French, and Spanish. Free of charge to customers.

d. Feed Additive Compendium. A guide to use of drugs in medicated animal feeds. The Miller Publishing Company, 12400 Whitewater Drive, Minnetonka, MN 55343. Revised annually with FDA approved changes.

C. Acknowledgements.

The authors wish to gratefully acknowledge assistance of numerous individuals who have offered information and made suggestions on materials presented. Special mention should be made of:

Dr. John P. Bishop, 11806b State Route 347, Marysville, OH 43040.

Dr. Ahmed Dardiri, USDA Plum Island Disease Center, Greenport, NY 11944.

Dr. David J. Farrell, The University of New England, Department of Biochemistry, Microbiology and Nutrition, Armidale, N.W.S. 2351, Australia.

Dr. J. David French, Poultry Technical Consultant, Lilly Research Laboratory, 210 Collier Drive, Albertville, AL 35950.

Drs. Henry L. Fuller, Leo S. Jensen, Hardy M. Edwards, Jr., Department of Poultry Science, The University of Georgia, Athens, GA 30602.

Dr. D. Earl Goodman, Turbeville, SC 29162.

Dr. Maxcy P. Nolan, Jr., Department of Entomology, The University of Georgia, Athens, GA 30602.

Permission was granted to reproduce illustrations by the 1) Peace Corps Information Service from Practical Poultry Raising; 2) Solvay Animal Health, Mendota Heights, MN, 55120 for diagrams from Salsbury Manual of Poultry Disease; 3) Will Graves 1985 Raising Poultry Successfully, Williamson Publishing, Charlotte, VT 05445; 4) Broiler Breeders Feeding and Management, Arbor Acres Farm, Inc. Glastonbury, CT 06033; and 5) original drawings by Dr. Erik L. Johnson, The University of Georgia, College of Veterinary Medicine, Class of 1991; and 6) Figures were re-drawn by Erin C. Pesti.

D. Authors.

W. MALCOLM REID (1910-1990)
Professor Emeritus of Poultry Science (Parasitology)
University of Georgia, Athens, Georgia 30602
He saw a need for a book dealing with basic poultry production and recruited others to help him develop one. He knew first-hand the value of poultry as a source of food and manure for soil enrichment.

GENE M. PESTI
Professor of Poultry Science and Animal Nutrition
Department of Poultry Science
University of Georgia, Athens, GA 30602
Telephone number: (706)542-1351 FAX number: (706)542-1827
B.S.A., 1972, The Ohio State University
M.S., 1976, Auburn University
Ph.D., 1980, University of Wisconsin-Madison
Dr. Pesti teaches poultry husbandry, economic management, and nutrition to undergraduate and post-graduate students. His research activity includes some 100 papers related to nutrition and management of poultry, and he has authored bulletins and computer software for undergraduate instruction. In 1987 he was a Visiting Research Fellow at the University of New England in Armidale, N.S.W., Australia, and in 1993 was a visiting scientist at INRA-Tours.

M. A. HAMMARLUND
Consulting Poultry Veterinarian
P. O. Box 7698
Riverside, CA 92513
Telephone (909) 687-2373
B.S., 1953, Kansas State University
D.V.M., Kansas State University
Dr. Hammarlund is a practicing veterinarian in Riverside, California with special interest in poultry. Before locating in California he was employed at the Ralston Purina Co. in St. Louis, Missouri, doing poultry disease research. Dr. Hammarlund held a "shuttle" missionary appointment to Bolivia in 1989 with the Christian Veterinary Mission.

PRAN VOHRA
Professor Emeritus
Department of Avian Sciences
University of California, Davis, CA 95616
Telephone (916) 752-3533
B.Sc. (Hons. School in Chemistry), 1940. Panjab University
M.Sc., 1942. Panjab University
Ph.D., 1958. University of California, Davis

Dr. Vohra taught courses in poultry nutrition, comparative nutrition of avian species, and livestock and poultry production in developing countries. He has worked in many countries, and as a consultant for U.S. Feed Grains Council, and for the American Soybean Association in China.

B.M. HARGIS
Professor, Department of Poultry Science
University of Arkansas
Fayetteville, AR 72701
Phone: (501) 575-4390 Fax: (501) 575-4202
E-mail: *bhargis@comp.uark.edu*
B.S., 1980, University of Minnesota
M.S., 1983, University of Georgia
D.V.M., 1986, University of Minnesota
Ph.D., 1987, University of Minnesota
From 1987-2000, Dr. Hargis was a faculty member at Texas A&M University and presently serves as director of the Poultry Health Research Laboratory at the University of Arkansas. He teaches poultry health to undergraduate and graduate students and is involved in research in a number of areas related to poultry health and production efficiency. Dr. Hargis, a diplomate of the American College of Poultry Veterinarians, maintains a research program in the area of poultry health and antemortem food safety intervention. Dr. Hargis has advised or co-advised more than 50 M.S. and Ph.D. students and has published numerous manuscripts and book chapters relating to food safety, poultry health and poultry physiology. Dr. Hargis is a past president of the Southern Poultry Science Society and in 1999 was inducted as a member of the Mexican National Veternary Medical Academy. His laboratory has been recognized by several awards including the Carrington Laboratories Research Award (1991), the Poultry Science Association Research Award (1993), USDA Certificate of Merit for Scientific Leadership (1994), the Texas Veterinary Medical Association Faculty Research Award (1994), the National Broiler Council Research Award (1998), and the Texas A&M University Vice Chancellor's Award in Excellence for Research (2000).

RANDLE W. MOORE
Veterinary Clinical Associate
Texas A&M University, College Station, TX 77483
Telephone number: (979) 845-5184
B.S., 1989, Texas A&M University
M.S., 1993, Texas A&M University
D.V.M., 1996, Texas A&M University
Ph.D., 2000, Texas A&M University

Dr. Moore, a diplomate of the American College of Poultry Veterinarians, teaches poultry health in the undergraduate and veterinary programs at Texas A&M University, where he currently serves as a Veterinary Clinical Associate. Previous poultry experience includes poultry judging and mentoring in 4H and FFA programs for high school students. Dr. Moore was recognized for his research and presentations at both Poultry Science Association and Southern Poultry Science Society meetings during his graduate studies. Research interests include several areas of poultry physiology and pathophysiology.

WILLIAM F. DEAN
Duck Specialist/Consultant
Former Researcher and Director
Cornell University Duck Research Laboratory
P.O. Box 217, 192 Old Country Road
Eastport, JY 11941
Before retirement, Dr. Dean was active in poultry research for over 40 years. His early work with H. M. Scott was on the amino acid requirements of the chick, and the development of a chemically defined diet that supported normal chick growth for use in poultry research. Subsequently he spent approximately 30 years at the duck laboratory on Long Island conducting research on nutrition and management of ducks. He is co-author with M. L. Scott on the book Nutrition and Management of Ducks. Dr. Dean has served as a consultant on duck matters for the American Soybean Association in China and for the American Feed Grains Council in Taiwan.

Christian Veterinary Mission (Publisher of this book)

Our vision is to see

Christ's love expressed through veterinary medicine.

Our mission is to

challenge, empower and facilitate veterinarians to serve through their profession, living out their Christian faith.

CVM also provides education and encouragement for those who desire to minister through service, prayer, relationship building, and modeling Christ's love.

About CVM

Christian Veterinary Mission (CVM) is a ministry of CRISTA Ministries, a registered non-profit Christian Service Organization 501(c)(3) based in Seattle, Washington, U.S.A.

CVM was founded in 1976 by Dr. Leroy Dorminy who came to realize the impact that veterinarians could have by integrating their faith with their practice, both locally and around the world. In 2008, CVM had nearly 30 veterinary professionals serving full-time internationally and over 200 veterinary professionals and student volunteers serve on short-term cross-cultural mission trips annually. CVM sponsors fellowship & prayer breakfasts at over 20 U.S. veterinary meetings each year and reaches out to veterinary students through Christian Veterinary Fellowship (CVF) groups in every veterinary school in the U.S. by encouraging them in spiritual growth and professional development.

There are over 3,500 veterinarians affiliated with CVM in the U.S. CVM also partners with organizations and networks in other countries that are focused on empowering Christian veterinarians. CVM has a volunteer advisory board of veterinarians who guide its vision, mission, and programming.

CVM books and the free International Animal Health Newsletter were written with small farmers, veterinarians, and agricultural development workers in mind. Our desire is that they would help individuals and groups develop an appropriate livestock program to meet community needs. CVM's Endowment Fund was started in the early years of the organization's life. The fund provides for meaningful programs that could not be funded by the regular budgeting process.

35303338R00080

Made in the USA
Lexington, KY
08 September 2014